Food Safety: Theory and Practice

Food Safety: Theory and Practice

Dennis Freeman

MURPHY & MOORE
www.murphy-moorepublishing.com

Published by Murphy & Moore Publishing,
1 Rockefeller Plaza,
New York City, NY 10020, USA
www.murphy-moorepublishing.com

Food Safety: Theory and Practice
Dennis Freeman

International Standard Book Number: 978-1-63987-232-9 (Hardback)

Cataloging-in-Publication Data

Food safety : theory and practice / Dennis Freeman.
p. cm.
Includes bibliographical references and index.
ISBN 978-1-63987-232-9
1. Food--Safety measures. 2. Food. 3. Food industry and trade--Quality control.
4. Foodborne diseases--Prevention. I. Freeman, Dennis.
TX531 .F66 2022
363.192 6--dc23

TABLE OF CONTENTS

PREFACE

The branch of science which studies the methods related to the storage, preparation, transportation and handling of food in ways which prevent food borne diseases is known as food safety. Some of the important practices of this discipline are related to the origin of food, labeling of food, food additives and food hygiene. According to the nature of contaminants, food contamination can be classified into physical contamination, chemical contamination and biological contamination. Physical contaminants include foreign bodies such as hair, paper, pieces of metal and plastic. Chemical contamination refers to spoilage of food products due to presence of substances like pesticides, herbicides, veterinary drugs, etc. Presence of harmful microorganisms including bacteria, viruses and parasites in food products is termed as biological contamination. This book unfolds the innovative aspects of food safety which will be crucial for the holistic understanding of the subject matter. Different approaches, evaluations and methodologies on food safety have been included herein. Those in search of information to further their knowledge will be greatly assisted by this book.

A foreword of all chapters of the book is provided below:

Chapter 1 - Food safety refers to preparing, handling, and storing food in a manner which prevents foodborne illnesses. Some of the hazards which are studied within food safety are physical hazards, chemical hazards and biological hazards. This chapter will briefly introduce all these significant aspects of food safety; **Chapter 2** - Food quality refers to the sum of all the properties and assessable attributes of a food item. Food quality evaluation and food quality control are two of its major aspects. This chapter closely examines these key concepts of food quality to provide an extensive understanding of the subject; **Chapter 3** - Food preservation seeks to prevent the growth of microorganisms and in turn slow the oxidation of fats which can cause the food to become rancid. Some of the methods used for the preservation of food are refrigeration, canning, curing and pickling, and smoking. The topics elaborated in this chapter will help in gaining a better perspective about these methods of food preservation; **Chapter 4** - There are numerous techniques which are employed to ensure that the food is safe for consumption. A few such techniques are food irradiation, food extrusion and vacuum packing. The topics elaborated in this chapter will help in gaining a better perspective about these techniques related to food safety; **Chapter 5** - The corruption of food by a physical, biological or chemical substance is termed as food contamination. It can cause food allergy as well as foodborne illnesses such as diarrhea, gastroenteritis, nausea and vomiting. This chapter discusses in detail these effects of food contamination.

At the end, I would like to thank all the people associated with this book devoting their precious time and providing their valuable contributions to this book. I would also like to express my gratitude to my fellow colleagues who encouraged me throughout the process.

Dennis Freeman

Chapter 1

Food Safety: An Introduction

Food safety refers to preparing, handling, and storing food in a manner which prevents foodborne illnesses. Some of the hazards which are studied within food safety are physical hazards, chemical hazards and biological hazards. This chapter will briefly introduce all these significant aspects of food safety.

Food Safety refers to handling, preparing and storing food in a way to best reduce the risk of individuals becoming sick from foodborne illnesses. Food safety is a global concern that covers a variety of different areas of everyday life. The principles of food safety aim to prevent food from becoming contaminated and causing food poisoning. This is achieved through a variety of different avenues, some of which are:

- Properly cleaning and sanitizing all surfaces, equipment and utensils.
- Maintaining a high level of personal hygiene, especially hand-washing.
- Storing, chilling and heating food correctly with regards to temperature, environment and equipment.
- Implementing effective pest control.
- Comprehending food allergies, food poisoning and food intolerance.

Regardless of why you are handling food, whether as part of your job or cooking at home, it is essential to always apply the proper food safety principles. Any number of potential food hazards exist in a food handling environment, many of which carry with them serious consequences.

Food Contamination

Food contamination refers to the ways that food has been depraved biologically, physically, or chemically. The contaminant could inter the food unintentionally pending agricultural produc-tion, environment, storage, transportation, sale and processing. In general, two sources of contamination are outside sources and formation in food that refer to primary and secondary contamination respectively. The main criteria for contamination judgment could be potential risk and the effect it has on human health. In this regard, mycotoxins and other microbial toxins, toxic elements, radioactive isotopes, nitroso compound, polycyclic hydrocarbon aromatic, halogen con-taining organic compounds, pesticides residues, veterinary drug residues, etc., are major critical food contaminants.

Mycotoxins produced by filamentous microfungi that can cause many diseases in vertebrate animals via ingestion, absorbtion (through the skin) and inhalation routes. Mycotoxins have been found in a variety of food commodities due to the mycotoxin producer fungus are able to grow on a vast range of foods. The most pronounced contamination has been initiated from the agricultural fields during several harvesting stages including preharvest, harvest, and postharvest. Mycotoxin contaminations finally continue to the consumer table. Poor management in all stages not only can lead to rapid deterioration in nutritional value but also provides proper condition for fungal growth and also mycotoxin production. The most efficient way to control food contaminants is the implementation of Good Manufacturing Practices (GMPs) and Hazard Analysis and Critical Control Point (HACCP) that will help prevent hazards in life. New approaches are based on identification of critical control point in production/processing of food that obtain optimum condition for mycotoxin production. To minimize and control mycotoxins in food chain all environmental and climate factors must be assessed.

Biological Contamination

Biological contamination generally realizes as contamination of food or environment with microorganisms and their derivatives such as toxins. In this regard, bacteria, viruses, fungi, and parasites are potential contaminants. They are found in food, walls, water, air, clothes, etc. The biological contamination also can occur via macroscopic organisms including rodents and insects. The biological contaminants cause human diseases via three mechanisms including infection, intoxication and immunologic responses.

Bacteria

Bacteria are small microorganisms that can grow in an ideal condition. They split and multiply so quickly. Harmful bacteria, called pathogen, are recognized as hazards in safety of food. Therefore the spread and incidence of them must be controlled in food. The common sources for bacterial growth and further distribution are the air, human body, dust, pets and pests, row food (meat, milk, vegetable, etc.), soil, kitchen/factory instruments, food handlers and cloths/hands. The extrinsic factors that provide optimum conditions for bacteria to survive include food (especially protein), water (water activity), oxygen, temperature and pH level. The control of these factors can result in well preservation of food. Table is illustrates major bacteria and their risks.

Table: The major bacteria and their risks.

Major bacteria	Risk contamination
Clostridium botulinum	Intoxication, even death
Listeria monocytogenes	Infection
Salmonella spp. (typhimurium, enteriditis)	Infection
Enterohaemorrhagic Escherichia coli	Infection
Campylobacter jejuni	Infection
Yersinia enterocolitica	Infection
Listeria monocytogenes	Infection
Bacillus anthracis	Infection
Bacillus cereus	Intoxication

Staphylococcus aureus	Intoxication
Clostridium perfringens	Infection
Vibrio spp. (vulnificus, parahaemolyticus)	Infection
Brucella abortus, B. suis	Infection
Shigella spp. (dysenteriae)	Infection

Viruses

Viruses are very tiny organisms that can grow and survive only in a host cell. They are able to enter food and water due to poor hygienic conditions. Viruses can also be found in people who disrespect hygienic practices. Viruses can only multiply and grow inside a living cell. They are very resistant to heat, drying, freezing, radiation, etc. and are also able to survive for a long time in food or environment. Viruses can enter food during processing, transportation through person to person contact. The awareness about the importance of viruses as food contaminants would result in good hygiene practices done by consumers to minimize the transmission of viral illnesses. Table shows the major viruses and their risk.

Table: The major viruses and their risk.

Major viruses	Risk contamination
Hepatitis A virus	Fever, abdominal discomfort.
Norwalk virus	Nausea, vomiting, diarrhea, abdominal pain (gastroenteritis), headache, low-grade fever.
Rotavirus	Vomiting, watery diarrhea, fever, abdominal pain.

Parasites

Parasite including worms and protozoa can enter food or water. They can also infect people through these ways. They need ambient environment and proper hosts to survive. Contamination of food occurs by fecal due to poor personal hygiene of food handler, improper disposal of human feces, improper sewage treatment and utilization of untreated sewage for crop culturing. Table shows major Parasites and their risks.

Table: The major parasites and their risks.

Major parasites	Risk contamination
Giardia lamblia	Diarrhea, abdominal cramps, fatigue, nausea, flatulence (intestinal gas), weightloss.
Entamoeba histolytica	Dysentery (severe, bloody diarrhea).
Ascaris lumbricoides	Intestinal, lung infection.
Diphyllobothrium latum	Attaches to intestinal wall.
Cryptosporidia	Respiratory, gastrointestinal illness.
Trichinella spiralis	Intestinal wall, enter the blood (to feed on it) and lymphatic system.
Toxoplasma gondii	Neurological disorders, particularly schizophrenia, bipolar disorder.
Taenia solium	Attaches to intestinal wall.
Anisakis spp.	Anisakiasis.

Chemical Contamination

Chemical contaminations of food are another human concern that involves numerous substances such as: agrochemicals, veterinary medicines residues, pesticides residues, environmental contaminants, heavy metals, persistent organic pollutants, and natural toxins; which happen in food during chemical reactions at processing such as acrylamide, furan, and heterocyclic amines. Other processes leading to the formation of contaminants include fermentation (e.g., ethyl carbamate, 3-monochloropropanediol) and disinfection (e.g., trihalomethanes). On the other hand, food contact materials are also kinds of chemical sources (e.g., formaldehyde, melamine, phthalates, and primary aromatic amines) that are able to leach into food. Some chemicals are naturally present in the environment, which includes ubiquitous pollutants such as dioxins and heavy metals may be increased by anthropogenic activity. Some common sources of chemical contamination are:

- Cleaning agents,
- Unwashed fruits and vegetables,
- Food containers made from non-safe polymers,
- Pest control products,
- Chemicals used in equipment maintenance.

Physical Contamination

Physical contamination refers to food that is contaminated by a foreign object during production process stages. Physical contaminants in food could come from external sources, (such as metal fragments), or internal sources (such as bone particles and pits). They can enter food accidentally during pre- and postharvesting due to poor agricultural practices and also in manufacturing, storage, transportation, or retail. Some physical contaminants are considered as food safety concerns such as glass. Sometimes a physical contaminated food can also be biologically contaminated such as the presence of a fingernail. Table provides a summary of common sources of physical contaminants in foods.

Table: Common sources of physical contaminants in foods.

Sources	Contaminants
Field	Rocks/stones/sand, asphalt, metals/bullets, concrete particles, bones, wood fragments and thorns.
Processing	Glass, ceramic/chards, metal fragments, staples, blades, clips, needles, keys, screws, magnet fragments, washers, bolts, screening, plastics, grease/lubricants, rubber, insulation/seal materials, nail polish, jewelry, coins, pieces of gloves, finger cots, bandages, cigarette butts, gum, bones, pits, fruit stones, nut and animal shells, medications/tablets/capsules, wood, pens, and pencils, rodents and insects.
Storage and distribution	Metal, plastic, and wood fragments, insects and rodents.

Cross-contamination

Cross-contamination occurs in food due to the contact of a contaminated substance coming from another food via many different ways and various sources including: Dirty clothes, Utensils, Coughing, sneezing or even touching the face and hair with food, Pests, Flies, cockroaches, mice and rats, Contacting Raw food, Garbage and waste materials.

Mycotoxin Contamination Problem

Mycotoxins are secondary metabolites that are produced by molds. Mycotoxins belong to biological contamination category. The majority of fungi can produce mycotoxins yet this potential is species specification. Several adverse effects on humans, animals, and crops originate from mycotoxins. The contamination of food with mycotoxins is a worldwide concern. Incidence of mycotoxins depends on temperature and humidity of a region that is prone to the fungal growth. The exposure to mycotoxins could take place by ingestion and or dermal and inhalation ways without involving the producer fungi. In fact these fungal toxins are a kind of abiotic hazard originated from biotic ones. The disease caused by mycotoxins is mycotoxicoses also resulted in different acute and chronic effects. Generally, the contaminants could enter the food unintentionally by agricultural production, environment, storage, transportation, sale and processing. Mycotoxins are natural contaminants contribute the food chain. The mycotoxin may contaminate the food during several stages of food chain from the soil to the plate. As a matter of fact mycotoxin contamination can occur in food by infection of crops not only when directly consumed by human but also consumed as feed. However ingested mycotoxin could result in its accumulation in body organs that enter food/feed through agricultural products, meat, milk or eggs. Various foods such as cereals, nuts, spices, fruits and also their products have a potent to be contaminated with mycotoxins at high content. Table shows major mycotoxin and producer microorganism.

Table: The major mycotoxins and producer microorganism.

Mycotoxin name	Producer microorganism
Aflatoxins	A. flavus, A. parasiticus, Aspergillus bombycis, Aspergillus ochraceoroseus, Aspergillus nomius, and Aspergillus pseudotamari
Ochratoxins (ochratoxin A)	Aspergillus ochraceus, Aspergillus carbonarius, Aspergillus melleus, Aspergillus sclerotiorum, Aspergillus sulphureus, Pichia verrucossum
Trichothecenes	Fusarium spp.
Zearalenone	Fusarium, F. culmorum, F. graminearum, F. sporotrichioides
Fumonisins	Fusarium proliferatum, Fusarium verticillioides
Tremorgenic toxins	Penicillium
Ergot alkaloids	Claviceps
Moniliformin	Fusarium species (mainly F. proliferatum)

The foods could contaminate with fungal toxins from farm at post- and preharvest stages. The implementation of Good Agriculture Practice and Good Manufacturing Practice are efficient strategies in preventing of mycotoxin contamination. However, every negligence in this field could provide proper condition for fungal growth and proliferation as well as Aspergillus growth and aflatoxin production in nuts. Therefore, the condition of production, processing, drying, handling, storage, transportation, and marketing must be controlled. It should be considered that further mycotoxin increase is difficult if the food products are preserved or stored under preventive conditions for fungal growth and mycotoxin production especially regarding water activity and temperature. The awareness of all people either producer or consumer about the ways of mycotoxin entering the food, the main stages of food chain involving in mycotoxin contamination, the optimum condition for increase of mycotoxin contamination risk and critical control tips in this field are necessary to the prevention of mycotoxin contamination.

Agricultural Production

Mycotoxins contamination of agricultural commodities can initiate from field and obtaining of conditions that conduct the fungal growth. The crops can be infected by molds at any line in the field. The production of mycotoxins due to mold growth is commonly associated with variation in weather conditions, plant stress, and humidity also inadequate feeding conditions. In this regard there are three main stages that develop the mycotoxin contamination in food staff from agricultural aspect. These stages include:

Pre-harvest

Mycotoxins can be produced in some natural food products due to the plants that are infected by mycotoxigenic species of molds group from farm. Pre-harvest practices include obtaining proper planting conditions such as soil ingredient, field qualification, crop rotation, irrigation, insect prevention, and antifungal treatment. As a matter of fact the pre-harvest condition control is the first line of mycotoxin prevention therefore the implementation of good agricultural practices (GAP) is needful. Some important tips for preventing mycotoxins in pre-harvest stage are listed below:

- Control of climate,
- Control of fertilizer,
- Control of insect, rodent, and birds,
- Control of weed,
- Biological control,
- Control of planting date control,
- Control of irrigation time,
- Prevention of early splitting in nuts,
- Control of water activity and water stress.

Harvest

Mycotoxin production could intensify by any inconsideration in harvest level. The main strategies in harvesting are including utilization of efficient harvesting/collecting/transportation equipment, attention to harvesting time, control of moisture and full maturity of product, inhibition of crop damages during harvesting. The delayed harvest could influence the development of mycotoxin contamination. The aflatoxin incidence in maize and nuts during delayed harvest were reported in many literatures. In the harvest stage it should be avoid contacting the harvested crop with the ground in order to prevent further contamination. Some of major tips in order to prevention of mycotoxins in harvest stage are listed below:

- Control of date of harvest,
- Control of the last irrigation,
- Control of storage condition,
- Control of blending of harvested products,
- Control of the contact with soil/ground,
- Control of harvest equipment hygiene,
- Control of damaging of grain,
- Sorting of defectives.

Post-harvest

Mycotoxin contamination more likely could occur in the post-harvest stage due to improperly handling. The high humidity during post-harvest prone the dry seed to absorb moisture followed by increase in water activity that conductive to contamination. Also the combination of temperature with moisture results in the extent of mycotoxin contamination risk. At the pos-tharvest stage, the observation of hygienic in all practices is necessary for safety guarantee of food products for example the use of clean transport vehicle free of any fungal growth. Consequently some tree nuts with high risk of mycotoxin contamination should be transferred to the processing plant as soon as possible after full maturity approximately within 6 h. The time of harvesting strictly have been recommended influence in mycotoxin production. Some crops when left in massive volume on the farm for a long time may present high level of mycotoxins. Some of important tips in post-harvest stage in order to prevention of mycotoxins are listed below:

- Rapid dehulling of crops (if it is needed),
- Rapid drying of crops (if it is needed),
- Separation of early splitting grain,
- Control of moisture content of product,
- Control of time and temperature during processing,
- Control of temperature and relative humidity during storage,
- Control of hygienic condition in all process,
- Control of additional water content of crops after washing,
- Control of packaging condition,
- Control of chemical preservatives.

Environment and Climate

Environmental conditions could increase the susceptibility of crops to infection by molds also favor fungal growth consequently mycotoxin production. Wounding of crops like tree nuts by birds, mammals, insects, may eventuate significant fungal infections. On the other hand, some insects carry mycotoxin (like aflatoxin) producing fungi associated with mycotoxin contamination in crops. The environmental factors affecting this contamination such as variation of seasons, disposal close to farm regions and the rates of insect population are all influenced by climate. Climate condition may directly influence some agricultural product by developing structural changes in crop. In this regard the hull cracking in nuts such as pistachio so called "early split" favor the fungal growth species especially Aspergillus spp. The rainfall that occurs at harvesting time may accelerate the fungal growth of crops. However in some geographical regions the time of high raining and high relative humidity of weather should be considered at harvesting and storage time of crops.

Storage

To avoid further mycotoxin contamination the agricultural product should be dried or de hulled immediately. The moisture content of stored food products must be lower than critical moisture (15% moisture and preferably to <13%) content according to water activity need for fungal growth (generally less than 0.7 at 25 °C). In this situation the competition of microorganism for water not only prevents further growth of fungi but also inhibits the mycotoxin production. Therefore in storage stage the main preventive action must be to decrease the moisture content and also the temperature. In Storage, any migration of moisture, condensation of moisture, and leaks should not occur.

The production of aflatoxin is strongly influenced by water activity of food commodities at storage time. On the other hand the infected food commodities specially crops and also nuts are able to provide adequate inoculums for incidence of the fungus to sound ones during poor storage practices. Storage management is essential in preventing fungal proliferation and mycotoxin formation in any harvested products.

Transportation

For many foods may face mycotoxin problems, the transportation conditions and time are great factors controlling the increase of mycotoxins content. During the transportation, some extrinsic factors like moisture content, relative humidity, temperature and hygienic control (cleanliness, insect control, etc.) could directly affect the safety of food. It is much emphasized that the transportation of foods under high humidity may result in mycotoxins increase. Transportation must be done in controlled conditions and any failure in this part may lead to decay of high volume of commodities.

Processing

Since mycotoxins are chemical and thermal resistant, they can be stable during heat, physical and chemical processing of food so, the prevention of mycotoxin production in row food is a critical control point in food production chain. In terms of food safety it should be considered that most treatment of foods such as roasting, boiling, pasteurization, irradiation, freezing, drying, blanching, exhausting, boiling, curing, foaming, frying are not effective on elimination nor reduction of mycotoxin in contaminated food. In this regard, it is better that all reduction or preventive strategies be performed before processing of food. On the other hand, these fungal toxins can also enter the human body via contaminated animal products (e.g., meat, egg, milk) due to feeding with mycotoxin contaminated feeds.

Prevention of Mycotoxins

According to numerous reports about high occurrence of mycotoxins in foods/feeds they are a constant concern worldwide. Although the mycotoxin producer molds spores are present all over the environment and related toxins can be formed on crops during harvest stages, Storage, processing. Also the mold spores are present in soil and plant debris able to infect growing agriculture products simply and fast at any point of handling. However, mycotoxin contaminations cover most of economic costs including the practices of prevention and mitigation, the reduced volume of contaminated foods, animal feed contamination and reduction in animal performance or health effects. Nowadays management of mycotoxins involves all actions of prevention, regulation, control, monitoring, tracing, avoidance, decontamination, detoxification and animal treatments. Even at such total management there may be levels of mycotoxin in food products unavoidably as a continual concern.

The most efficient tool for mycotoxin problems is the prevention of mold growth in fields especially during postharvest practices. Additionally, environmental factors can immensely affect the production of mycotoxin by fungal species. In terms of predictive proceeding the predictive models have been developed as decision supporting systems to plan proper crop protection strategies in fields. Innovative detection and diagnostic tools are also available to monitor the occurrence of mycotoxigenic fungi in fields and after harvest. When contamination is not prevented, several approaches can be employed to help remove mycotoxins from the contaminated commodities, including physical, chemical, and biological techniques. Detoxification processes should destroy or inactivate mycotoxins, by guaranteeing the nutritional value of food. Research is needed to study the fate of mycotoxins during decontamination, detoxification, and food processing. A holistic approach should be adopted to monitor, prevent, and control mycotoxigenic fungi and mycotoxins in food products.

Foodborne Illnesses

Foodborne illness, also called foodborne disease is any sickness that is caused by the consumption of foods or beverages that are contaminated with certain infectious or noninfectious agents. Most cases of foodborne illness are caused by agents such as bacteria, viruses, or parasites. Other agents include mycotoxins (fungal toxins), marine biotoxins, and the toxins occurring in poisonous mushrooms; metals such as lead, mercury, and cadmium, which may contaminate food through air, water, or soil pollution; organic pollutants, such as dioxin and polychlorinated biphenyls (PCBs), which are by-products of some industrial processes; and prions (abnormal forms of normally harmless proteins). The agents of foodborne illnesses cause a variety of conditions, ranging from gastroenteritis to reproductive or developmental disorders and neurodegenerative diseases, such as bovine spongiform encephalopathy (mad cow disease). People sometimes refer to foodborne illnesses, particularly those involving gastroenteritis, as food poisoning. Food poisoning, however, is a type of foodborne illness and specifically is caused by toxins present in foods; typically, those toxins are produced by bacteria and cause symptoms soon after the contaminated food is consumed. By contrast, certain other types of foodborne illnesses arise from ingestion of the infectious organisms or chemicals themselves and may take days to develop.

Although the incidence of foodborne illness is difficult to estimate, since many cases go unreported, the burden of illness is thought to be high. For example, tens of millions of people worldwide are affected annually by diarrheal diseases, a common proxy measure for foodborne illness. Foodborne illnesses often occur as outbreaks that have the potential to affect large numbers of people. For instance, in 1988 in China an outbreak of hepatitis A, caused by the consumption of contaminated clams, affected more than 300,000 people, and an outbreak of salmonellosis in the United States in 1994, caused by the consumption of contaminated ice cream, affected 224,000 people. Foodborne illnesses can be deadly. In 1985, for example, a listeriosis outbreak in California, involving a contaminated cheese product, caused 48 deaths out of 142 cases.

Common Foodborne Illnesses

There are a number of different types of foodborne illnesses. Escherichia coli is a type of bacteria that lives in the intestines of many animals, including humans and cattle. Most strains of E. coli are not harmful to humans. However, several strains can have serious health effects when ingested by humans. For example, E. coli 0157:H7, which lives in the intestines of cattle, is usually ingested in undercooked ground beef, although it may also be transmitted through unpasteurized milk and fruit juice, contaminated water, uncooked produce, and person-to-person contact. Symptoms of E. coli 0157:H7 infection often include severe abdominal cramps and bloody diarrhea. In a small percentage of cases, hemolytic uremic syndrome may develop, which can result in kidney failure and death. Examples of other strains that cause illness in humans include E. coli O104:H4, O26:H11, and O111:H8.

Salmonellosis is caused by Salmonella bacteria, which commonly are found in the intestines of mammals, reptiles, and birds and usually are spread to humans through the consumption of foods of animal origin, including eggs, meat, and milk. Symptoms of salmonellosis include fever, headache, abdominal pain, nausea, vomiting, and diarrhea; in persons with weakened immune systems or poor health, it can be life-threatening.

Listeriosis, caused by the bacterium Listeria monocytogenes, is most often transmitted through milk, soft cheeses and ice cream, raw vegetables, and raw meat and poultry. Because L. monocytogenes can grow at low temperatures, certain foods that are refrigerated for long periods of time are particularly likely routes of transmission. Symptoms of listeriosis include flulike illness, with fever, fatigue, and muscle aches. Some people experience a loss of balance, confusion, or even convulsions. Listeriosis is dangerous particularly for pregnant women, because it can cause miscarriage and stillbirth, and for infants and persons with weakened immune systems, in whom it can cause meningitis (inflammation of the membranes that cover the brain and spinal cord) or septicemia (blood poisoning).

Campylobacteriosis is caused by Campylobacter bacteria. In some countries it is more common than salmonellosis, and worldwide it is the most commonly identified bacterial cause of diarrheal illness. Campylobacteriosis is transmitted mainly through drinking water, undercooked poultry, and raw milk. Because Campylobacter bacteria live in the intestines of healthy birds, most raw poultry can be assumed to be contaminated with it. The symptoms of campylobacteriosis include fever, nausea, severe abdominal pain, and diarrhea. Major health consequences, such as neurological conditions and reactive arthritis, may develop in a small number of cases.

Prevention and Control

Major foodborne illnesses, including salmonellosis and E. coli infections, are reportable diseases in many places, meaning that infections caused by those agents must be reported by physicians and medical laboratories to local, state, or national health departments. However, because most cases of foodborne illness are mild and are not diagnosed, the reported number of cases is assumed to be an undercount of the true number of cases.

Many laws regulating the production, transport, and preparation of food are intended to prevent foodborne illness and limit its consequences. For example, laws have been implemented to help prevent the contamination of raw food, to mandate its safe preparation and storage, and, if necessary, to close restaurants or food suppliers responsible for disease outbreaks or who fail to follow safe food-hygiene practices. There are many means by which raw food may be contaminated, including irrigating or washing with unclean water, contamination of meat and poultry with fecal matter during the slaughtering and packaging processes, preparation by food handlers who carry bacteria or viruses on their hands, or using utensils and preparation surfaces that are not clean.

Cooking at a sufficient temperature kills many microbes and parasites. However, some microbes may also be present in cooked food, such as when contamination occurs during handling after cooking or pasteurization. In some instances, contamination post-cooking or post-pasteurization may not pose a health risk, if only a small number of microbes are present. Most bacteria grow rapidly at room temperature, whereas refrigeration or freezing keeps them from multiplying (L. monocytogenes is a notable exception). Hence, even cooked foods must be promptly refrigerated in order to prevent the multiplication of disease-causing organisms.

Food Preservation

Food preservation is one of the methods to protect food from unwanted microbial growth. After the food is produced, we store and protect by covering the rice and curry with lids to keep away flies and other insects. By this, we are protecting it from any infection caused by them. This is a short-term condition. Food preservation, on the other hand, is done to preserve food for a longer time.

Objectives of Food Preservation

Following are the important objectives of food preservation:

1. To prevent microbial contamination.
2. To kill pathogens.
3. To minimize food spoilage and food poisoning.

Food Preservation Methods

Food preservation started long back in ancient times. Cooling, freezing, fermentation, sun-drying, etc., are few age-old food preservation techniques. With the advent of technology, modern methods of food preservation were developed. Chemicals and other natural substances were used for preservation. These substances are known as preservatives.

Methods of Food Preservation.

Chemical Method

Salt and edible oils are two main preservatives which are used since ages to prevent microbial growth. This is why we add extra oil to pickles. Preservation by salt is known as salting. Salting helps to preserve fruits for a long term. Meats and fishes can also be preserved by salting. Other synthetic preservatives include vinegar, sodium benzoate, sodium metabisulphite, etc.

- Sugar: Sugar is another common preservative used in jams and jellies. Sugar is a good moisture absorbent. By reducing moisture content, it restrains the microbial growth.
- Heat and Cold Methods: Boiling and refrigeration prevent around 70 percent of microbial growth. Boiling kills the microorganisms that cannot tolerate extreme temperatures. Thus, it helps in food preservation. Refrigerators have very low temperatures. Since microbes do not get optimum temperature they need for growth, their growth is inhibited. Pasteurization developed by Louis Pasteur is used until today to preserve milk.
- Smoking: Smoking prevents dehydration in fish and meat and thus prevents spoilage. The wood smoke contains a large number of anti-microbial compounds that slow the rancidification of animal fats.
- Canning: The food contents are sealed in an airtight container at high temperatures. Meat, fish, fruits are preserved by canning.
- Sterilization: This method is carried out to remove microbes from food. For eg., milk sterilization at 100 °C kills the microbes.
- Dehydration: It is the process of removal of water from food. It is the simplest method and prevents food spoilage by removing water.
- Lyophilization: This is the process of freezing and dehydration of the frozen product under vacuum.
- Radiation: This method is also known as cold sterilization. The UV rays, X rays, gamma

rays kill all the unwanted microbes present in food. The principle behind all these methods is either to reduce the temperature conditions required for microbial growth or to evaporate the water content.

Food Processing

Food processing is the set of methods and techniques used to transform raw ingredients into food or food into other forms for consumption by humans or animals either in the home or by the food processing industry. Food processing typically takes clean, harvested crops or slaughtered and butchered animal products and uses these to produce attractive, marketable, and often long-life food products. Similar processes are used to produce animal feed. Extreme examples of food processing include the expert removal of toxic portions of the fugu fish or preparing space food for consumption under zero gravity.

Examples of some processed foods.

The benefits of food processing include the preservation, distribution, and marketing of food, protection from pathogenic microbes and toxic substances, year-round availability of many food items, and ease of preparation by the consumer. On the other hand, food processing can lower the nutritional value of foods, and processed foods may include additives (such as colorings, flavorings and preservatives) that may have adverse health effects.

Food Processing Methods

Beer fermenting at a brewery.

Common food processing techniques include:

- Removal of unwanted outer layers, such as potato peeling or the skinning of peaches.
- Chopping or slicing, such as to produce diced carrots.
- Mincing and macerating.
- Liquefaction, such as to produce fruit juice.
- Fermentation, as in beer breweries.
- Emulsification.
- Cooking, by methods such as baking, boiling, broiling, frying, steaming, or grilling.
- Mixing.
- Addition of gas such as air entrainment for bread or gasification of soft drinks.
- Proofing.
- Spray drying.
- Pasteurization.
- Packaging.

Performance Parameters for Food Processing

When designing processes for the food industry, the following performance parameters may be taken into account:

- Hygiene, measured, for instance, by the number of microorganisms per ml of finished product.
- Energy consumption, measured, for instance, by "ton of steam per ton of sugar produced".
- Minimization of waste, measured, for instance, by the "percentage of peeling loss during the peeling of potatoes".
- Labor used, measured, for instance, by the "number of working hours per ton of finished product".
- Minimization of cleaning stops, measured, for instance, by the "number of hours between cleaning stops".

Benefits

More and more people live in the cities far away from where food is grown and produced. In many families, the adults are work from home and therefore there is little time for the preparation of food based on fresh ingredients. The food industry offers products that fulfill many different needs: from peeled potatoes that simply need to be boiled at home to fully pre-pared ready meals that can be heated up in the microwave oven in a few minutes.

Microwave oven

Benefits of food processing include toxin removal, preservation, easing marketing and distribution tasks, and increasing food consistency. In addition, it increases seasonal availability of many foods, enables transportation of delicate perishable foods across long distances, and makes many kinds of foods safe to eat by de-activating spoilage and pathogenic micro-organisms. Modern supermarkets would not be feasible without modern food processing techniques, long voyages would not be possible, and military campaigns would be significantly more difficult and costly to execute.

Modern food processing also improves the quality of life for allergy sufferers, diabetics, and other people who cannot consume some common food elements. Food processing can also add extra nutrients such as vitamins.

Processed foods are often less susceptible to early spoilage than fresh foods, and are better suited for long distance transportation from the source to the consumer. Fresh materials, such as fresh produce and raw meats, are more likely to harbor pathogenic microorganisms (for example, Salmonella) capable of causing serious illnesses.

Drawbacks

In general, fresh food that has not been processed other than by washing and simple kitchen preparation, may be expected to contain a higher proportion of naturally occurring vitamins, fiber and minerals than the equivalent product processed by the food industry. Vitamin C, for example, is destroyed by heat and therefore canned fruits have a lower content of vitamin C than fresh ones.

Tate & Lyle brand Corn Syrup being moved by tank car.

Food processing can lower the nutritional value of foods. Processed foods tend to include food additives, such as flavorings and texture enhancing agents, which may have little or no nutritive value, and some may be unhealthy. Some preservatives added or created during processing, such as nitrites or sulfites, may cause adverse health effects.

Processed foods often have a higher ratio of calories to other essential nutrients than unprocessed foods, a phenomenon referred to as "empty calories." Most junk foods are processed, and fit this category. High quality and hygiene standards must be maintained to ensure consumer safety, and failure to maintain adequate standards can have serious health consequences. Processing food is a very costly process, thus increasing the prices of foods products.

Trends in Modern Food Processing

Health

- Reduction of fat content in final product, for example, by using baking instead of deep-frying in the production of potato chips.
- Maintaining the natural taste of the product, for example, by using less artificial sweetener.

Hygiene

The rigorous application of industry and government endorsed standards to minimize possible risk and hazards. In the U.S., the standard adopted is HACCP.

Efficiency

- Rising energy costs lead to increasing usage of energy-saving technologies, for example, frequency converters on electrical drives, heat insulation of factory buildings, and heated vessels, energy recovery systems.
- Factory automation systems (often Distributed control systems) reduce personnel costs and may lead to more stable production results.

Industries

Food processing industries and practices include the following:

- Cannery,
- Industrial rendering,
- Meat packing plant,
- Slaughterhouse,
- Sugar industry,
- Vegetable packing plant.

Food Fermentation

Fermentation could be described as a process in which microorganisms change the sensory (flavor, odor, etc.) and functional properties of a food to produce an end product that is desirable to the consumer.

Common Fermented Foods

A selection of the most common fermented foods that have wide geographical distributions are shown in Table. The key type of microorganisms associated with these foods is also included.

Table: Examples of the more Common Fermented Foods.

Food	Principal Ingredient	Key Microorganisms
Wine	Grapes	Yeasts
Beer	Barley	Yeasts
Cider	Apples	Yeasts
Sake	Rice	Molds
Bread	Wheat	Yeasts
Yogurt	Milk	LAB (lactic acid bacteria)
Cheese	Milk	LAB
Buttermilk	Milk	LAB
Kefir	Milk	LAB + yeasts
Vinegar	Grapes	Yeasts followed by Acetobacter or Gluconobacter
Tempeh	Soybeans	Molds
Soy sauce	Soybeans	Molds + LAB + yeasts
Pickled cucumbers	Cucumbers	LAB + yeasts
Sauerkraut	Cabbage	LAB

Fermentation as a Preservation Method

As new preservation techniques have been developed, the importance of fermentation processes for food preservation has declined. Yet fermentation can be effective at extending the shelf life of foods and can often be carried out with relatively inexpensive, basic equipment. Therefore, it remains a very appropriate method for use in developing countries and rural communities with limited facilities. In addition, the non-dependence of fermentation on the use of chemical additives to the food appeals to the "more aware" consumer market. The chemical composition of most foods is relatively stable; therefore, generally preservation is based on eliminating microorganisms or controlling their growth and the overall composition of the microflora. To reduce or prevent microbial spoilage of food, four basic principles can be applied:

- Minimize the level of microbial contamination onto the food, particularly from "high-risk" sources (asepsis).

- Inhibit the growth of the contaminating microflora.
- Kill the contaminating microorganisms.
- Remove the contaminating microorganisms.

Fermentations use a combination of the first three principles. Fermentations should not be expected to sterilize substandard raw products, but rather should use high-quality substrates. Microorganisms can improve their own competitiveness by changing the environment so that it becomes inhibitory or lethal to other organisms while stimulating their own growth, and this selection is the basis for preservation by fermentation. A number of different bacteriocidal and bacteriostatic factors that can be produced by lactic acid bacteria (LAB) are shown in Table. Fermentation improves the safety of foods by decreasing the risks of pathogens and toxins achieving the infective or toxigenic level, and extends the shelf life by inhibiting the growth of spoilage agents, which cause the sensory changes that make the food unacceptable to the consumer.

Factors produced by the metabolic activity of microorganisms that can contribute to the increased stability and safety of fermented foods:

- Low pH.
- Organic acids, e.g., lactic acid, acetic acid, and formic acid.
- Low redox potential.
- Nutrient depletion.
- Accumulation of inhibitors, e.g., toxins, bacteriocins, antibiotics, lactococcins, nisin, natamycin, hydrogen peroxide.
- Ethanol.
- Diacetyl.
- Carbon dioxide.

Microbial Contamination of Foods

Foods are derived from other living organisms and during their development and preparation they are continuously exposed to microbial contamination. The resultant contaminating microflora can have different effects on the food. These include negative effects such as spoilage, where the food becomes unfit for human consumption or health risks when infectious or toxigenic microorganisms are present. Negligible effects on the food occur when the microflora does not cause disease or any detectable changes in the food. However, benefits can also be reaped from the action of the microorganisms when their activity brings about improvements in the appeal of the food. In developed countries, the improved appeal is the major reason for microbial fermentations of foods continuing today.

The nutrient content and intrinsic properties of many raw foods make them ideal environments for microbial replication. The rate at which the microorganisms grow depends not only on the intrinsic properties of the food (pH, redox potential, water activity, etc.) but also on the conditions under which it is being stored, the extrinsic factors, for example, temperature. Therefore, many raw food types need to be consumed soon after production to be of high nutritional value. Without preservation measures, delays lead to the nutrients being degraded and utilized by the contaminating microflora.

A major consideration needs to be that under ideal conditions microorganisms can grow very rapidly, being able to double in number in a short period of time. It must also be noted that there is a variation in the optimum environmental conditions for different types and species of microorganisms, for example, microorganisms can be categorized into broad groups such as aerobes and anaerobes depending on their tolerance and use of oxygen and psychrophiles, mesophiles, and thermophiles based on the temperature range optimum for their growth. In addition, the biochemical activity of different microorganisms varies and may change in response to fluctuations in environmental factors, leading to a range of metabolic end products. By manipulating the environmental conditions, it is possible to select for specific kinds of microorganisms that impart a particular taste, odor, texture, or appearance to the food. This is the basis of fermentation.

Table: Examples of microbial metabolic end products used in fermented foods.

Metabolic End Product	Example of Uses
Carbon dioxide	Leavening bread
Ethanol	Alcoholic beverages
Acetic acid	Vinegar
Lactic acid	Fermented vegetables
Flavor compound - Diacetyl	Dairy products
Flavor compound - Acetaldehyde	Yogurt

Benefits of Fermented Foods

Microorganisms per se can be used as food sources, but in many instances it is their effects on other food sources that are of major interest. The acceptability of a food to the consumer is based mainly on its sensory properties. The sought-after sensory properties of fermented foods are brought about by the biochemical activity of microorganisms. Fermented foods were developed simultaneously by many cultures for two main reasons: (i) to preserve harvested or slaughtered products, which were abundant at certain times and scarce at others and (ii) to improve the sensory properties of an abundant or unappealing produce. However, a range of benefits can be obtained from food fermentations, some of which are shown in Table. Consequently, fermented foods and drinks still retain an important role in the human diet. Fermentation has low energy demands and can often be carried out without sophisticated technology and designated plants. The simple techniques mean that the procedures can often be carried out in the home. Also, a number of studies have shown that consumers regard fermented food products as healthy and natural, increasing consumer demands and their profitability.

Table: Potential benefits of fermented foods.

Increased	Lowered
Safety	Toxicity
Health benefits	Cooking time
Retail value	Production costs
Nutritional value	Equipment needs
Digestibility	Levels of anti-nutritional factors
Suitability for subsequent processing: • Sensory properties • Ease of storage and transportation • Shelf life	

Microorganisms Used in Food Fermentations

A variety of groups of microorganisms are frequently used in fermented foods. The principal groups are shown in table.

Table: Principal groups of Microorganisms used for food fermentations.

Microbial Group	Product
Lactic acid bacteria (LAB)	Lactic acid
Acetic acid bacteria	Acetic acid
Yeasts	Alcohol and carbondioxide
Molds	Enzymes

Lactic Acid Bacteria

LAB perform an essential role in the preservation and production of wholesome foods. Examples of lactic acid fermentations include (a) fermented vegetables such as sauerkraut, pickled cucumbers, radishes, carrots, and olives; (b) fermented milks such as yogurt, kefir, and cheeses; (c) fermented/ leavened breads such as sourdough breads; and (d) fermented sausages. LAB have been grouped together as they possess a range of common properties, and all produce lactic acid that can kill or inhibit many other microorganisms. The primary use of lactic acid in the food industry is as a preservative, an acidulant, or a dough conditioner. LAB are subdivided based on their products from glucose fermentation. Homofermenters produce lactic acid as the major or sole product from glucose, while heterofermenters produce equimolar amounts of lactate, carbon dioxide, and ethanol. Heterofermenters have an important role in producing aroma components such as acetaldehydes and diacetyl. LAB have a range of methods for outcompeting other microorganisms. Their most effective mechanism is to grow readily in most foods, producing acid, which lowers the pH rapidly to a point where other competing organisms can no longer grow. Lactobacilli also have the ability to produce hydrogen peroxide, which is inhibitory to spoilage organisms, while lactobacilli are relatively resistant to hydrogen peroxide. The role of hydrogen peroxide as a preservative is likely to be minor, especially when compared with acid production. Carbon dioxide produced by heterofermenters also has a preservative effect, resulting partially from its contribution to anaerobiosis.

Consumers are taking a greater interest in the quality of foods and are creating a demand for chemicalfree, "natural health" foods. This has stimulated extensive research into the applications of LAB for both the control of pathogenic and spoilage microorganisms and also for health promotion. A range of potential health benefits has been associated with the consumption of LAB. Some benefits are as a consequence of their growth and activity during food fermentations, and some from the resultant colonization of the gastrointestinal tract.

Table: Potential health benefits from lactic acid bacteria.

Benefits
From foods • Improved nutritional value, e.g., production of vitamins or essential amino acids. • Reduced toxicity, e.g., by degradation of noxious compounds. • Increased digestibility and assimilability of nutrients.
From colonization • Control of intestinal infections. • Improved digestion of lactose. • Inhibition of tumor growth. • Lowering of serum cholesterol levels. • Immune stimulation.

Acetic Acid Bacteria

A second group of bacteria with importance in food fermentations are the acetic acid producers. Acetic acid is one of the oldest chemicals known; it is named after the Latin word for vinegar "acetum." The acetic acid bacteria are acid tolerant, grow well at pH levels below pH 5.0, are Gram-negative, motile rods, and are obligate aerobes. They derive energy from the oxidation of ethanol to acetic acid following the reaction shown below.

$$C_2H_5OH + O_2 \rightarrow CH_3COOH + H_2O$$

They are found in nature where ethanol is produced from the fermentation of carbohydrates by yeasts, such as in plant nectars and damaged fruits. Other good sources are alcoholic beverages like fresh cider and unpasteurized beer. In liquids, they grow as a surface film because of their demand for oxygen. The acetic acid bacteria consist of two genera, Acetobacter and Gluconobacter. Acetobacter can eventually oxidize acetic acid to carbon dioxide and water using Krebs cycle enzymes referred to as overoxidation. This is not the case with Gluconobacter. The most desirable action of acetic acid bacteria is in the production of vinegar. The same reaction can also occur in wines, when oxygen is available, and here the oxidation of alcohol to acetic acid is an undesirable change, giving the wine a vinegary off-taste.

Yeasts

Yeasts are widely distributed in natural habitats that are nutritionally rich and high in carbohydrates, such as fruits and plant nectars. Yeasts are rarely toxic or pathogenic and are generally acceptable to consumers. After extensive study, yeasts have been classified into about 500 species. However, only a small number are regularly used to make alcoholic beverages. Saccharomyces cerevisiae is the most frequently used and many variants are available. Saccharomyces cerevisiae ferments glucose but does not ferment lactose or starch directly. Yeasts are used to produce ethanol, CO2, flavor, and aroma. The reaction can be represented by the following equation:

$$C_6H_{12}O_6 \rightarrow 2C_2H_5OH + 2CO_2$$

$$\text{yeast} + \text{glucose} \rightarrow \text{ethyl alcohol and carbon dioxide}$$

Other metabolic products include minor amounts of ethyl acetate, fusel alcohols (pentanol, isopentanol, and isobutanol), sulfur compounds, and leakage of amino acids and nucleotides that can all contribute to the sensory changes induced by yeasts.

Molds

The majority of fungal species have filamentous hyphae and are referred to as molds. They are grouped into four main classes based on the physiology and production methods of their spores. Molds are aerobic and have the greatest array of enzymes. Some molds are used in the food industry to produce specific enzymes such as amylases for use in bread making. They are relatively tolerant to extreme environments and are able to colonize and grow on most foods. Molds are important to the food industry, both as spoilers and preservers of foods and in particular in fermentations for flavor development. Certain molds produce antibiotics, while mycotoxin production by others is an emerging cause of concern in the food industry. The Aspergillus species are often responsible for undesirable changes in foods, although some species such as A. oryzae are used in fermentations of soybeans to make miso and soy sauce. Mucor and Rhizopus are also used in some traditional food fermentations. Rhizopus oligosporus is considered essential in the production of tempeh from soybeans. Molds from the genus Penicillium are associated with the ripening and distinctive flavor of cheeses.

Starter Cultures

Fermented foods may be produced by the action of fermentative microorganisms naturally found on the raw materials or in the production environment. However, to improve reliability "starter cultures" are frequently used. Starter cultures may be pure or mixed cultures. Using mixed starter cultures can reduce the risks of bacteriophage infection and improve the quality of the foods when the organisms are mutually beneficial. Food fermentations frequently involve a complex succession of microorganisms induced by dynamic environmental conditions. Fermentative microorganisms must be safe to eat even in high numbers and must produce substantial amounts of the desired end product(s). For practical reasons, the organisms should be easy to handle and should grow well, enabling them to outcompete undesirable microorganisms. The organism also needs to be genetically stable with consistent performance both during and between food batches. In many traditional fermentations, the natural micro flora were used for the fermentation. Natural fermentations have a degree of unpredictability, which may be unsatisfactory when a process is industrialized. Starter cultures are increasingly used to improve not only the reliability, but also the reproducibility and the rate at which the fermentation is initiated. Failed, poor-quality, or unsafe products lead to loss of customers and revenue, therefore their incidence must be minimized. The composition of starter cultures is based on knowledge of food-grade microbial genetics, metabolism, and physiology as well as their interactions with foods. Starter cultures are now developed mainly by design rather than by screening. The overall objective is to exploit the properties of the starter cultures to ensure reproducible standards of safety and quality.

Classification of Fermented Products

Fermented foods are classified in a number of different ways. They may be grouped based on the microorganisms, the biochemistry, or on the product type. Campbell-Platt identified seven groups for classification, namely, (1) beverages, (2) cereal products, (3) dairy products, (4) fish products, (5) fruit and vegetable products, (6) legumes, and (7) meat products, whereas Steinkraus classified fermentations according to the type of fermentation.

Fermented Products

Alcoholic Beverages

Throughout history alcoholic beverages have had a place in most cultures. They require the alcoholic fermentation of fruits or other high-sugar materials by yeasts. The alcohol content of the beverage acts as a preservative and many of these products have long shelf lives. Over the years, brewing yeasts have evolved by selection and mutations, and have been developed by genetic engineering. Major advances have been made in improving the characteristics of the fermentation strains driven by the high revenue associated with the alcoholic beverage industry.

Beer

Beer is produced by the fermentation of partially germinated cereal grains, referred to as malt, by yeasts. Beers have a final ethanol content of about 3%–8%; a huge variety of beers exists and they include ales, lagers, and stouts. Both lagers and ales can be either light or dark in appearance. Ale is produced using Saccharomyces cerevisiae, a top fermenter yeast, whereas lagers are produced using pure cultures of Saccharomyces carlsbergensi, a bottom fermenter yeast. Ales are produced using warm fermentation temperatures, 12 °C–18 °C and lager fermentation temperature is generally cold, 8 °C–12 °C. Most beer produced is of the lager variety.

Wine

Wine can be produced from any fruit juice with sufficient levels of fermentable sugars, in most cases wine is a beverage obtained by full or partial alcoholic fermentation of fresh, crushed grapes or grape juice (must), with an aging process. Wine-type grapes from cultivars of Vitis vinifera vines are most commonly used to produce wines.

Distilled Spirits

The fermented products discussed above can only produce a maximum alcohol content of about 17%. Concentrations in excess of this inhibit the metabolism of the yeasts. To obtain higher alcohol concentrations, the fermented product must be subsequently distilled. Whiskey, gin, vodka, rum, and liqueurs are examples of distilled spirits. Although the process for producing most products of these types is quite similar to that for beers, the content of alcohol in the final products is considerably higher.

Lactic Acid Products

Dairy Products

Yogurt is a coagulated milk product obtained by lactic acid fermentation through the action of Streptococcus thermophilus and Lb. delbrueckii subsp. bulgaricus. Yogurt is prepared using either whole or skim milk, where the nonfat milk solids are increased to 12%–15% by concentrating the milk, or adding powdered skim milk or condensed milk. The concentrated milk is pasteurized at 82 °C–93 °C for 30–60 min and cooled to the starter incubation temperature of 40 °C– 45 °C. Yogurt starter is then added at a level of around 2% by volume and incubated for 3–5 h, or until the titratable acidity of the final product reaches 0.85%–0.90% or a pH of 4.4–4.6. The yogurt is then cooled to 5 °C to inhibit further acid production. The type of yogurt starter used can change the physical characteristics of the final yogurt product.

Cheese is a concentrated milk product obtained after coagulation and whey separation of milk, cream or partially skimmed milk, buttermilk, or a mixture of these products. Cheese may be consumed fresh or after ripening. Cheese is commonly made from cow, ewe, goat, or buffalo milk. The majority of cheeses are made from pasteurized milk. The use of sub-pasteurization heat treatment of milk or thermization is also practiced to limit heat-induced changes in milk without compromising microbiological safety. There are over 400 varieties of cheeses representing fewer than 20 distinct types, and these are grouped or classified according to texture or moisture content, whether ripened or unripened, and if ripened, whether by bacteria or molds.

Fermented Vegetables

A large number of vegetables are preserved by lactic acid fermentation around the world. The most important commercially fermented vegetables are cabbage (sauerkraut), cucumbers, and olives. Others include carrots, cauliflower, celery, okra, onions, and peppers. Typically, these fermentations do not involve the use of starter cultures and rely on the natural flora. Brine solutions are prepared in the fermentation of sauerkraut, pickles, and olives. The concentration of salt in the brine ranges from 2.25% for sauerkraut to 10% for olives. The fermentation yields lactic acid as the major product. The salt extracts liquid from the vegetable, which serves as a substrate for the growth of LAB. Growth of undesirable spoilage microorganisms is restricted by the salt. Aerobic conditions should be maintained during fermentation to allow naturally occurring microorganisms to grow and produce enough lactic acid, and to prevent growth of spoilage microorganisms. Olives receive a special treatment before bringing in that green olives are treated with a 1.25%–2% lye solution (sodium hydroxide), usually at 21 °C–25

°C for 4–7 h. This treatment is necessary to remove some of the oleuropein, a bitter compound in olives. In some countries, the fermentation of cucumbers is controlled by the addition of acetic acid to prevent growth of spoilage microorganisms, buffered with sodium acetate or sodium hydroxide, and inoculated with Lb. plantarum alone or in association with Pediococcus cerevisiae.

Fermented Animal Products

The primary reason for developing methods to ferment meats and fish was to extend the shelf life of these highly prized, perishable foods. Gram-positive micrococci have an important role in these fermentations. Several products became popular, including fermented sausages, fish sauces, and fish pastes. Many of the traditional fermentation methods are still used although the primary reason for their use is no longer preservation, but because the products are popular for their enhanced flavors.

Fermented Sausages: A variety of procedures for producing stable, fermented meat sausages have developed around the world. In general, preservation of the meat is achieved by adding salts and the generation of lactic acid by bacteria, which leads to a rapid fall in the pH. Micrococci, staphylococci, and yeasts are responsible for the development of color, taste, and flavor during the fermentation. In addition to fermentation, sausage processing may also include curing, smoking, drying, and aging to improve both the flavor and the shelf life. In addition to the major inhibitory factors of low-water activity achieved by the addition of salt, and in some cases drying and the accumulation of lactic acid, a "hurdle effect" is created by a combination of other inhibitors, contributing to the preservation of sausages.

Fermented sausages are prepared by mixing ground meat with various combinations of spices, flavorings, salt, sugar, additives, and frequently, starter cultures. Common additives include acidulant, ascorbic acid, and colorings. Pork, beef, mutton, or turkey meat can be used, but to achieve good sensory properties and safety, the meat must be fresh and of high quality. The meat is generally used raw, with no heat processing, as this can damage the texture of the sausage product. Frequently fermented sausages are eaten without any cooking stage.

Fermentation of fish is most common in southeast Asia where fish is a major component of the human diet. The carbohydrate content of fish is low, usually less than 1%; therefore, for lactic fermentation an additional source of carbohydrate is required. Ingredients such as rice and garlic may be added as carbohydrate sources; the carbohydrate reserve in garlic is inulin. The higher the level of supplemented carbohydrate, the faster the fermentation. The product is often ready after only a few weeks, making the process much more efficient. The supplementation of carbohydrates enables the fish to ferment and an acidic, stable product to be made. The most common products from fish are produced by microbial fermentation and by the degrading activity of autolytic fish enzymes. These products are known as fish sauces and fish pastes.

Combined Fermentations

The release of carbon dioxide by microorganisms has two major roles in food fermentations: (i) it can act as a leavening agent and (ii) it can be used for carbonation of beverages. One of the most common uses of carbon dioxide is to leaven dough during bread making.

Bread

The use of yeasts to produce bread dates back thousands of years. Breads have relatively short shelf lives; therefore, the primary reason for their production was not preservation, but to improve the digestibility and eating appeal of grains. To make leavened bread, flour produced from grains, such as wheat, that contain gluten proteins are used. When the bread is kneaded, the gluten forms a matrix that makes the dough elastic and extensible. These properties enable the dough to stretch and retain sufficient amounts of gas to produce loaves with good volume and a fine, soft open structure. The gas is a combination of air incorporated during mixing and kneading and CO_2 produced from the fermentation of sugars by yeasts. The yeasts contribute to the flavor and provide an appealing aroma. When the loaves are baked the proteins are denatured, fixing the structure, and the low levels of ethanol produced by the yeasts evaporate.

Sourdough

Sourdough bakery products have an extended shelf life. The fermentation combines the metabolic activity of LAB for souring and yeasts for leavening. Methods for their fermentation date back thousands of years. The sourness of the product depends on many factors, including fermentation temperature and time, type of grain, and the strains of yeast and LAB. The complexity of the bread flavor is based on the lactic and acetic acids produced by LAB and flavor compounds formed by the activity of endogenous cereal enzymes, microbial metabolism, and the baking process. The metabolism of LAB and yeasts also provides a range of desirable aroma products.

Vinegar

Vinegar is one of the oldest known culinary products. It is thought that it was discovered by accident from spoilt wine, in fact it is named after the French term "vin aigre" meaning sour wine. Vinegar is classified as a condiment that contains a minimum of 4% w/v (40 g/L) acetic acid and has a pH value between 2.0 and 3.5. The strength of vinegars may also be quoted in grains, with 10 grains being equivalent to a concentration of 1% acetic acid. Higher strength vinegars may be used for pickling; spirit vinegar is made from an alcoholic solution that has been distilled. Although vinegar has been produced for thousands of years, it remains very popular; it is estimated that the annual worldwide production of vinegars is around 2000 million liters. Vinegar is one of the great successes of the preservation industry, although acetic acid has numerous applications in the food industry. The shelf life of a wide range of foods is extended by storing the product submerged in vinegar; this includes pickled vegetables such as gherkins, olives, and onions.

Vinegars are produced from two-stage fermentation: initially an anaerobic, alcoholic fermentation of sugars by yeasts, followed by oxidation of the ethanol to acetic acid by bacteria; this second reaction is known as acetification. Acetification is also a common cause of spoilage of alcoholic beverages. Acetification can be described by the equation:

$$C_2H_5OH + O_2 \rightleftharpoons CH_3COOH + H_2O$$

Kefir

Kefir is produced by an acid/alcohol fermentation of pasteurized milk with a mixture of LAB, yeasts, and other bacteria. The final product is acidic, slightly alcoholic, liquid to semiliquid, and effervescent, and is consumed as a beverage. Kefir grains are used to inoculate the milk. Kefir grains comprise proteins, polysaccharides, and a mixture of microorganisms, mainly lactosefermenting yeasts, and aroma bacteria and LAB. The yeasts consist mainly of Candida kefir and Saccharomyces kefir, while the LAB comprise mainly of Lactobacillus kefir, Leuconostoc species, and L. lactis. The yeasts are responsible for the production of ethanol and carbon dioxide from lactose, the lactococci produce lactate from lactose, and the lactobacilli and Leuconostoc species are responsible for the production of lactate, acetate/ethanol, and carbon dioxide. Kefir fermentation requires a moderate room temperature (17 °C–23 °C). The final composition of kefir includes 0.8% lactic acid, 1%–3% alcohol, diacetyl, and acetaldehyde.

Oriental Fermented Products

The production of soy sauce, miso, and saki involves koji fermentation. Koji comprises soybeans or grains on which molds grow to produce enzymes such as proteases, lipases, and amylases. The fungal enzymes produced digest proteins, carbohydrates, and lipids into nutrients that are used by microorganisms in subsequent fermentations. Koji is produced in many varieties depending on the products to be manufactured. Koji differs in terms of the molds, the substrate, the method of preparation, and the stage of harvest.

Soy Sauce

Soy sauce is a dark brown liquid produced by the fermentation of soybeans and wheat in a salt brine. The manufacture of soy sauce starts with the treatment of raw material. Soybeans or defatted soybean flakes are moistened and cooked. The cooked beans are then mixed with roasted, cracked wheat in varying ratios for each type of soy sauce. The mixture is inoculated with a pure culture of Aspergillus oryzae (A. soyae). After 3 days of fermentation, 17%–19% salt solution is added to the koji to produce a mash called moroni. LAB such as Pediococcus soyae or Lb. delbrueckii are allowed to grow on the moroni to make it acidic enough to prevent spoilage and to make it acidic in taste. Yeasts such as Saccharomyces rouxii and Torulopsis sp. grow on the moroni to produce alcohol and help the formation of flavor. The moroni is aged, pressed to produce a liquid, soy sauce, which is then pasteurized.

Tempeh

Tempeh is a protein-rich food that is considered one of the world's first meat analogs. It is made by growing the mold R. oligosporus or related species on soaked, dehulled, partially cooked soybeans, knitting them into a firm cake, which can be sliced and deep-fried or cut into cubes and used in place of meat in soups. Tempeh production is not a means of improving shelf life, but it does improve the acceptability and the nutritional quality of its raw material (soybeans).

Food Hazards

Food safety is a major public health problem. It not only causes mortality and morbidity but also affects food trade. Food can be become unsafe on account of various food hazards and they are classified into three categories based on the nature of hazards. They are physical, chemical and biological hazards.

Physical Hazards

Physical hazards are either foreign materials unintentionally introduced to food products (ex: metal fragments in ground meat) or naturally occurring objects (ex: bones in fish) that are hazardous to the consumer. A physical hazard contaminates a food product at any stage of production. Food processors should take adequate measures to avoid physical hazards in food.

Physical Hazards in Food

- Hard or sharp objects are potential physical hazards and can cause:
    - Cuts to the mouth or throat.
    - Damage to the intestines.
    - Damage to teeth or gums.
- The presence of physical hazards in food can trigger a food recall, affecting the brand name of your company and product.

Factors Determining a Potential Risk

Factors that cause potential risk to consumers in food products include:

- Size: Health Canada states that anything in a food product, that is extraneous, and measures two millimeters or more in size can be a health risk.
- Type of consumer: Products that target infants, the elderly, etc. have a higher risk level.
- Type of product: The form the product takes such as infant formulas, beverages, etc. can increase risk level.
- Physical characteristics: hardness, shape and sharpness of a product can affect risk level.

Common Physical Hazards

Common sources of physical hazards in food include:

- Glass: light bulbs, glass containers and glass food containers.
- Metal: fragments from equipment such as splinters, blades, needles, utensils, staples, etc.
- Plastics: material used for packaging, fragments of utensils used for cleaning equipment.
- Stones: incorporated in field crops, such as peas and beans, during harvesting.
- Wood: splinters from wood structures and wooden pallets used to store or transport ingredients or food products.

- Natural components of food: hard or sharp parts of a food (ex: shells in nut products) if consumers do not expect them.

Preventing Common Physical Hazards

There are many ways food processors can prevent physical hazards in food products. Assess every step of your operation for potential sources of contamination:

- Inspect raw materials and food ingredients for field contaminants, such as stones in cereals that were not found during receiving.
- Handle food according to Good Manufacturing Practices (GMPs). (Example: avoid inclusion of physical hazards such as jewellery or false fingernails in food products by using proper personnel practices.)
- Eliminate potential sources of physical hazards in processing and storage areas. (Example: use protective acrylic bulbs or lamp covers to prevent contamination by breakable glass.)
- Install an effective detection and elimination system for physical hazards. (Example: metal detectors or magnets will detect metal fragments in the production line while filters or screens will remove foreign objects at the receiving point.)
- Establish an effective maintenance program for the equipment in your facility to avoid sources of physical hazards such as foreign materials that can come from worn out equipment.

Detecting and Eliminating Physical Hazards

There are several methods available to detect foreign bodies on food processing production lines:

- Magnets can be used to attract and remove metal from products.
- Metal detectors can detect metal in food and should be set up to reject products if metal is detected. Equipment should be properly maintained to ensure it is always accurate and doesn't produce false positives.
- X-Ray machines can be used to identify hazards such as stones, bones and hard plastics, as well as metal.
- Food radar systems transmit low-power microwaves through food products to identify foreign bodies such as metals, plastics, bones or kernels in food.

Chemical Hazards

A chemical hazard is any substance that can cause a health problem when ingested or inhaled. They include toxins, dangerous chemicals, residue of excess chemicals used in processing food products. If your facility follows Good Manufacturing Practices (GMPs), you can prevent chemical hazards.

It is difficult to establish cause and effect relationship especially with chemical hazards in food as the effects may not be acute. Excessive exposure to chemicals is suspected to be responsible for a range of human health problems. A wide variety of diseases like cancer, kidney and liver dysfunction, immune system suppression, musculoskeletal disease, birth defects, premature births, impeded nervous and sensory system development, reproductive disorders, mental health problems, cardiovascular diseases, genitor-urinary disease, old-age dementia and learning disabilities. It is important to know that what are the different types of chemical hazards, their sources in food, so that effective control measures can be taken to make food safe. To understand chemical harzad, firstly we have to know about:

- Type and nature of chemical hazards in food.
- Source of chemicals and control measure.

Chemicals can be present in food due to various reasons. They may come in to food due to air, water and soil pollution or as they are used in food crops and food animals or some chemicals are formed during the food processing or some of them are added to food for various technological functions during food processing. Toxins produced by various microorganisms are also considered as chemical contaminants. Depending on the nature and source of chemicals, they are broadly classified in to the following classes:

- Mycotoxins.
- Marine toxins.
- Pesticide residues.
- Veterinary Drug Residues.
- Environmental contaminants.
- Food additives.
- Process contaminants.
- Food contact materials.
- Cleaning agents and sanitizers.

Mycotoxins

These are toxins produced by certain fungi. According to one estimate more than 400 different mycotoxins are discovered but very few of them of concern from food safety point of view. Field fungi present on field food crops and storage fungi under favourable conditions produce toxins. Each of the toxin has different health effects in humans. Among them 5 major groups of mycotoxins which are often found in food and feed. They are:

- Aflatoxins.
- Deoxynivalenol/nivalenol.
- Zearalenone.

- Ochratoxin.
- Patulin.

1. Aflatoxins: Aflatoxins are classified as potent human carcinogens by International Agency for Research on Cancer. They are produced by Aspergillus flavus and Aspergillus parasiticus. There are five different aflatoxins ie Aflatoxin B1, B2, G1, G2 and another found in Milk as M1 Maize, Groundnut, Chillies are the major foods where aflatoxin contamination, except Aflatoxin M1 is often found. Aflatoxin M1 is metabolite of Aflatoxin B1 which comes in to milk if the milch cattle's feed is contaminated with aflatoxin B1.

2. Deoxynivalenol/nevalenol: These mycotoxins affect human health causing acute temporary nausea, vomiting, diarrhea, abdominal pain, headache, dizziness, and fever. Some species of Fusarium produce these toxins and it generally infects maize, wheat, oats, barley, rice, and other grains in the field or during storage.

3. Zearalenone: Zearalenone is a mycotoxin produced by several Fusarium species and is found in maize wheat, barley, sorghum and rye. zearalenone is primarily a field contaminant, toxin production may occur under poor storage conditions. No human health effects have been reported but it has been shown to be having oestrogenic activity in experimental animals.

4. Ochratoxin A: Ochratoxin A is a mycotoxin produced by several fungal species including Aspergillus ochraceus, A. carbonarius, A. niger, and Penicillium verrucosum. It is generally found in dry beans, maize, wheat ,barley, oats and green coffee beans. Ochratoxin A causes nephrotoxicity and renal tumors in the experimental animals. Balkan endemic nephropathy, chronic interstitial nephropathy and other renal diseases have been linked to Ochratoxin A exposure.

5. Patulin: Patulin is produced by several species of Aspergillus and Penicillium. Patulin is found in unfermented apple and grape juice and field crops. There are no studies to indicate the human health effects of patulin. However earlier there was a concern that it could be a carcinogen but the International Agency for Research on Cancer concluded that there is insufficient evidence to consider it as human or animal carcinogen. The other toxic effects observed in animal experiments are gastrointestinal lesions, distension and haemorrhage in the stomach and small intestine.

Marine Toxins

As the name indicates marine toxins are chemicals found in sea food. They enter sea food due to infection or bioaccumulation due to ingestion of toxic algae. The most common food poisonings due to marine toxins are:

- Paralytic shellfish poisoning.
- Scombrotoxic fish poisoning.
- Ciguatera poisoning.
- Neurotoxic shellfish poisoning.
- Amnesic shellfish poisoning.

1. Paralytic shellfish poisoning: Consumption of seafood like mussel, oysters, crabs. lobsters and clams have caused the paralytic shellfish poisoning and the typical symptoms include headache, dizziness, nausea and muscular coordination problems.

2. Scombrotoxic fish poisoning: It is also called histamine poisoning is caused due to consumption of contaminated seafood like tuna and mackerel. The symptoms include rash, flushing, abdominal pain vomiting and diarrhea.

3. Ciguatera poisoning: This poisoning is caused by consumption of one particular fish called Barracuda which feeds on contaminated fish. This fish is generally found in the eastern side of Atlantic Ocean, Red Sea and Caribbean Sea. Symptoms include nausea, vomiting, cramps and diarrhoea.

4. Neurotoxic shellfish poisoning: It is caused by consumption of oysters, clams and mussels which contain certain neurotoxin. The symptoms of this poisoning are numbness, tingling sensation and lack of limb coordination.

5. Amnesic shellfish poisoning: It is very rare poisoning caused by a toxin made by a diatom called Nitzchia pungens. Ingestion of these diatoms by mussels and subsequently such seafood consumption by human causes this poisoning. Symptoms include dizziness, headache, disorientation, and permanent short-term memory loss. In severe poisoning, seizures, focal weakness or paralysis, and death may occur.

Pesticide residues: A large number of pesticide are used to control pests in food crops (includes feeds for food animals). They leave a residues in food are called pesticide residues. More scientific definition is provided by World Health Organsation. As per that "Any substance or mixture of substances in food for man or animals resulting from the use of a pesticide and includes any specified derivatives, such as degradation and conversion products, metabolites, reaction products, and impurities that are considered to be of toxicological significance." Their presence in the food is inevitable but the levels of residues present should not have any health effects on the consumers. Hence regulatory bodies identify Maximum Residue Limits (MRLs) for each of the pesticide in each of the food, feed and commodities.

These are the upper legal levels of a concentration for pesticide residues in or on food or feed based on good agricultural practices and to ensure the lowest possible consumer exposure. Toxicity data is generated on each of the pesticide using experimental animals, so it is difficult to explain all the toxic effects of all pesticides. Presence of Pesticide residues in foods beyond their respective MRLs is cause of concern for the consumers. Each country has its own list of MRLs for pesticides, hence it is necessary to understand the country's requirement while processing food.

Veterinary Drug Residues

Food animals and fish like humans are also susceptible for infections and diseases. Veterinary Drugs are used for their treatment or cure the diseases, leave residues in edible animal product like meat, milk, fish eggs etc. Residues of veterinary drugs include the parent compounds and/or their metabolites in any edible portion of the animal product and include associated impurities of the veterinary drug concerned. Like MRLs for pesticides, Veterinary Drugs also have MRLs for edible animal products. Maximum Residue Limit (MRL) is the maximum concentration of residue resulting from the use of a veterinary drug (expressed in mg/kg or ug/kg on a fresh weight basis) as per good veterinary practice. The MRL also takes into account other relative public health risks as well as food technological aspects. All the necessary toxicological data is generated for each of veterinary drug before it is permitted to be used. Presence of veterinary drug residues beyond their MRLs is a public health concern as they may result in antibiotic resistance in consumers or allergy in sensitive individuals. Some of the hormone residues may act as endocrine disruptors. Permitted veterinary drugs and MRLs may also vary from country to country.

Environmental Contaminants

The environmental contaminants refers to toxic chemicals present in soil, air and water. These chemicals may be from natural sources or due to industrial pollution. Environmental contaminants include toxic metals, dioxins, polycyclic aromatic hydrocarbons(PAHs) and polychlorinated biphenyls.

Toxic metals: The following four metals are generally discussed the context of food contamination:

- Lead
- Cadmium
- Mercury
- Arsenic

1. Lead: Lead occurs naturally in rock and soil and also used in many industries. Lead is ubiquitous in environment and hence present in air, water and soil as well as in food. Lead can enter the food chain through various routes. For example, plants uptake lead from the soil and it can also be deposited on food surfaces. Food animals are exposed to Lead, through their feed and water. The concentrations of Lead found in food are low but now it is well established even at lower levels of lead is known to produce damage to various organs which on long run causes loss of cognition, shortening of attention span, alteration of behaviour, dyslexia(know as reading disorder, it is a learning disability), attention deficit disorder, hypertension, renal impairment, immunotoxicity and toxicity to the reproductive organs. It is very difficult to eliminate Lead from food, hence to reduce the risk having any health effect on consumers, tolerance limits have been prescribed for the foods.

2. Cadmium: Unlike Lead, Cadmium is relatively rare metal found nature. It is also used in many industries including fertilisers. Cadmium can enter food through uptake by plants or food animals exposed to contaminated feed & water or seafood as molluscs or crustaceans. The amount of cadmium in food is likely to be very low but chronic exposure at lower levels of cadmium can lead to a build-up of cadmium in the kidneys and high build-up will damage the kidneys. Chronic exposure will also make bones fragile.

3. Mercury: Mercury is also an industrial pollutant and it is released in to environment by power plants, cement plants, certain chemical industries and it is also used to manufacture thermometers, thermostats and automotive light switches etc. Once mercury enters water due to environmental pollution, it is converted to methyl mercury by bacteria, the methyl mercury consumed by fish, and in fish it bio accumulates. Consumption of such contaminated fish causes mercury

poisoning. High risk population for mercury exposure is pregnant woman and children. Methyl mercury is neurotoxin and it affects brain development of children, developing foetus in pregnant woman. In adults it causes memory loss and tremors.

4. Arsenic: Arsenic is naturally present in earth's crust thus it is found in air, soil and water as environmental contaminant. In some countries high levels of arsenic is present in ground water. States of Bihar, Jharkhand, Chattisgarh, Tripura, West Bengal, Assam have high levels of arsenic in ground water. Human exposure of arsenic through food is essentially due to use of contaminated water for drinking, irrigation of food crops and industries. The long term effects of arsenic ingestion through food contamination leads to initially pigmentation of skin and later skin cancer. In addition, it can also cause lung and urinary bladder cancer.

Dioxins

Dioxins are class of chemicals formed during incineration of waste , forest fires and industrial processes like paper pulp bleaching and herbicide manufacturing. Among all the dioxins, the most toxic chemical in the class is 2, 3, 7, 8-tetrachlorodibenzo-para-dioxin (TCDD). Dioxins accumulate in fatty tissues of food animals and thus consumers are exposed to dioxins. Health effects like risk of cancer, developmental problems, heart disease, diabetes etc have been documented mostly in population occupationally exposed to dioxins.

Polycyclic Aromatic Hydrocarbons (PAHs)

These are another set of chemicals formed due to burning of organic matter. Not all PAHs are toxic but most toxic and researched PAH is Benzopyrine. PAHs are found environment due to vehicular emissions or industrial pollution. PAHs are also formed during the smoking of fish or meat or barbequing etc. They are considered both as environmental and process contaminants. Data from experiments show them to be mutagenic, genotoxic and carcinogenic.

Polychlorinated Biphenyls (PCBs)

PCBs are group of compounds which are used in production of various products like diffusion pump oils, transformers, capacitors, hydraulic and lubricating fluids, flame retardants, paints, adhesives, Plasticizers etc. They end up as environmental contaminants. Animal experiments indicated that PCBs can cause hormonal imbalance, reproductive failures and cancer. Their production is banned in many countries due to their highly toxic nature. Fish found in contaminated site or food animals grazing in contaminated feed will accumulate PCBs in their fatty tissues. People are exposed to PCBs through consumption of contaminated fish or animal fat.

Food Additives

Food additives are chemicals intentionally added to food for a technological purpose in the manufacture, processing, preparation, treatment, packing, packaging, transport etc. These additives are permitted by regulatory agencies and they are specific to food and quantity is also fixed for the specific technological function. All permitted food additives are safe at the levels of their permitted levels of use. However, it is the misuse i.e. use of unpermitted additives or use of permitted additives beyond their permitted quantities, that creates food safety concern. Some of the colour additives like Rhodamine, Malachite green, metanil yellow or sudan dyes are not permitted to be used in foods as they are not safe. Commonly used preservatives like Sulphites may aggravate asthmatic attacks in sensitive individuals. Like in the case of pesticide residues or veterinary drug residues, the list of permitted additives may vary from country to country.

Process Contaminants

These are the chemicals formed during food processing and their presence in the food is a cause of food safety concern. The following are some of the important process induced contaminants:

- Acrylamide
- Furan
- Nitrosamines

1. Acrylamide: Acrylamide is a chemical formed during processing of starch rich food material a high temperature more than 120c with low moisture. It was detected in potato chips, French fries, bread, biscuits and coffee. Acrylamide is a carcinogen hence a food safety concern.

2. Furan: Furan is also a chemical formed preparation of food during heat treatment (sterilising, roasting, baking) from natural components of the food. It is found in coffee, canned foods, baked foods etc. Furan is found induce cancer in experimental animals and International Agency for Research on Cancer(IARC) has declared it as probable human carcinogen.

3. Nitrosamines: Nitrite is a food additive used in meat and meat products as preservative and colour fixative. Nitrites react with break down product of protein and produce nitrosamines. They are also formed during processing like in bacon and beer. Most of the nitrosamines are mutagenic and can become carcinogens under certain conditions. Nitrosamines are generally detected in cured meats.

Food Contact Materials

All those materials which are likely to come in contact with food, like packaging materials i.e, plastics, paper, metal and machinery, utensils, containers etc. Chemicals from these food contact material can migrate into food and make it unsafe. There has been a concern on the type of plastics used for food packaging, as some of them are known to migrate in to food and are carcinogens. Polyvenyl choride (PVC) and polystyrene two unsafe plastics. Bisphenol A which is used for polycarbonate bottles is suspected to be endocrine disruptor. Machinery used for food processing should not result in contaminating the food. Hence it is generally advised to use stainless steel material. Even the containers, utensils used for food materials also should not leach any toxic material in to food. Specifications for food grade/approved food contact materials is provided by the food regulatory agencies in each country.

Cleaning Agents

Cleaning is a prerequisite for effective sanitization. Organic matter from food residues such as oils, grease and protein needs to be removed from the premises or machinery or utensils for sanitization to take place. Cleaning agents like detergents, acids, abrasives used are likely to leave residues. These residues should not become health hazards, hence only approved cleaning agents are permitted to be used in food processing.

Sanitizers

Sanitization follows cleaning. Heat or chemicals are used for sanitization. Sanitization removes almost all pathogenic organisms. Chemicals like antiseptics, disinfectants, bactericides or bacteriostatic agents are used in the food industry for sanitization purpose. Here again the residues should not become health hazards. Only sanitizers which can be used in the food processing are identified based on their safety.

Control of chemical hazards: A wide variety of chemical hazards are likely to be present in the foods and they can enter the food chain any stage from production to consumption. Once they enter food chain, it is difficult to eliminate them. Prevention is the better method of their control. Pesticide residues and veterinary drug residues needs to be controlled at production stage by following Good Agriculture Practices and Good Veterinary Practices. Mycotoxins can be effectively controlled by proper drying of food grains. Putting in place, proper quality control mechanism for in coming raw materials to control marine toxins, environmental contaminants, food contact materials will help. Good manufacturing practices for control of food additives, process contaminants, cleaning and sanitizers. Finally application of food hygiene basic principles at all stages of food chain will help in minimizing both chemical and biological hazards.

Biological Hazards

Any biological substances in or condition of food with the potential to cause an adverse health humans is called biological hazard. Biological hazard can include pathogenic strains of bacteria, helminths, protozoa, algae, toxin or virus that affects human health. The presence of biological hazards in food causes foodborne disease. The foodborne disease continues to be a common and serious threat to public health all over the world. It is a major cause of morbidity and mortality.

Types of Biological Hazards

The most common biological hazards are mainly categorized in to following types:

- Bacteria: Bacteria are microscopic living organisms and are omnipresent. They are characterized by lack of distinct nucleus. Example: Salmonella, L.monocytogenes, Staphylococcus, Pathogenic E.coli etc.
- Virus: Viruses are ultramicroscopic, requires living host for multiplication. Virus contains either DNA or RNA as their nucleic acid. Example: Norwalk virus, Hepatitis E virus, Norovirus etc.
- Fungi: Fungi are saprophytic organisms that lack chlorophyll and dead/decaying organic matter or living organisms. Example: Aspergillus flavus, Penicillium verrucosum.
- Protozoa: Protozoa are unicellular, eukaryotic organisms Example: Toxoplasma gondii.
- Helminths: Parasitic, large multicellular organisms Example: Trematodes.

- Algae: large group of eukaryotic organisms. Example: Alexandrium sp., Gymnodinium sp., Dinophysis sp., and Pronocentrum sp.

Pathogenic Bacteria

Pathogenic Escherichia Coli

Escherichia coli are gram negative, non sporing rods, motile with peritrichous flagella. Most of the strains of E. coli are harmless and found naturally in the gut of humans and animals. Its presence in food is considered as fecal contamination of food or water. However particular pathogenic strains of E. coli causes traveler's diarrhea, harmorrhagic colitis, haemolytic uraemic syndrome etc. Based in the virulence genes acquired, different types of pathogenicity are conferred to certain strains of E.coli. These strains are classified as entero-pathogenic E.coli (EPEC), entero-toxigenic E.coli (ETEC), entero-invasive E.coli (EIEC), entero-hemorrhagic E.coli (EHEC), entero-aggregative E.coli (EAEC). The type of disease varies with the different virulence factors, haemolysin production, toxin production exhibited or expressed by the organism. A wide variety of foods have been implicated including milk, milk based sweets, non vegetarian foods like meat and fish etc. E.coli 0157:H7 was found in beef, milk, paneer and ice-cream but in UK it is more common in beef burgers, meat and dairy products. Control of E. coli in food production is done through effective control of raw materials, process conditions and post process contamination.

Salmonella

Salmonella are facultatively anaerobic, usually motile, Gram negative rods causes mild to severe food poisoning, typhoid and paratyphoid. Salmonella spp. has been considered as most important causal agent of foodborne illness throughout the world. Salmonella is mainly associated with raw meats, poultry and dairy products. However, many other foods have been implicated in outbreaks like kheer, fish, stale rice, frozen food etc. Salmonella was found to be more common in meat, poultry, seafood and khoa but in other country like UK, it is more common in meat, poultry, eggs and dairy products.

Number of 10 cells of salmonella have been responsible for causing foodborne disease. Young children, immune compromised people, sensitive individuals and elderly people are much more likely to become ill after consumption of low number of cells. Salmonella takes 12-72hrs for symptoms of the illness to occur. The main symptoms of Salmonella food poisoning is diarrhea, abdominal pain, vomiting and fever.

Salmonella can be readily destroyed in foods with a high water activity. In frozen foods or foods containing low water activity, Salmonella ahs shown to survive for many months, even years. The effective way of controlling or minimizing the hazard of Salmonella from foods involve control of reaw materials, personal, equipment and environmental hygiene, manufacturing process conditions, post process contamination, retail and catering practice and consumer handling.

Listeria Monocytogenes

L.monocytogenes is Gram positive foodborne pathogen causes listeriosis. This bacteria is widely distributed in the environment and occurs in almost all food raw materials. There are six distinguishable species of listeria are recognize, L. monocytogenes, L. innocula, L. welshimeri, L. seeligeri, L. ivanovii and L. grayi. The most commonly occurring species in food are L. innocula and L. monocytogenes. It is widely associated with foods like milk, sea foods and vegetables and its percentage of contamination varies from 1-12%.

Listeria food poisoning occurs by consumption of food containing very high levels of (>10^7 per gram) L.monocytogenes. The incubation time of L.monocytogenes is <24hrs and its symptoms are vomiting and diarrhea. L.monocytogenes is known to grow at the temperature used for refrigeration. The control of Listeria spp. in foods is dependent on preventing contamination of raw materials, preventing recontamination from the factory environment, equipment or personnel, minimizing its growth, if present during the shelf life of the final product.

Campylobacter

Campylobacter are Gram negative, micro-aerophilic, non sporing exhibits darting reciprocating motility. Campylobacter jejuni is the most significant species among the family campylobacteriaceae causing foodborne disease. It is usually present in the gut of wide range of animals. It is widely associated with the foods like poultry and sea foods. Quite small number of Campylobacter around 500-600 cells of bacteria can cause illness. The appearance of symptoms will take 2-7 days.

Contaminated water is one of the vehicles for the transmission of Campylobacter infection. Campylobacter does not survive as well as other pathogens such as Salmonella, although it has been shown to survive in foods for longer at lower temperatures. It is very vulnerable to factors such as high temperature and dry environments and also to the presence of oxygen in atmospheric conditions.

Staphylococcus Aureus

Staphylococcus aureus is Gram positive, catalase positive coccal shaped foodborne bacteria. Staphylococcus aureus produces golden yellow pigment on selective media. Staphylococcus aureus present as part of the normal microflora of humans and other animals. It is being carried on skin and nasal cavities of healthy human population. This bacteria is widely associated with foods like khoa, fish, meat and dairy products. Its percentage of contamination varied from 20-36%. Staphylococcal food poisoning is a type of foodborne intoxication. Staphylococcus aureus produces heat stable toxin called enterotoxin. Staphylococcal intoxication results from ingestion of enterotoxins synthesized during growth in foods. Staphylococcus aureus intoxication results from ingestion of foods containing immunologically distinct enterotoxins, A, B, C1, C2, C3, D, E, F, G, H and I. Most of the Staphylococcus aureus outbreak are caused due to SEA and SED enterotoxins.

Nausea, vomiting and diarrhea are the main symptoms of Staphylococcus food poisoning. Symptoms usually appear within 2-4 hrs of ingestion of implicated food. The minimum dose of Staphylococcus aureus enterotoxin is considered to be 1µg or less. Staphylococcus aureus is not heat resistant and it can be easily destroyed at temperature normally used for food processing. Enterotoxin produces by Staphylococcus aureus are heat stable and are not destroyed by temperature used in food processing.

Yersinia Enterocolitica

Yersinia enterocolitica is a Gram negative, rod shaped bacteria. This bacteria is having ability to grow at 40 °C. It causes foodborne disease called yersiniosis. This bacteria is widely associated with foods like milk, butter milk and lassi. Its percentage of contamination varied from 5-59%. The symptoms appear within 2-3 days and characterized by abdominal pain and diarrhea. Yersinia enterocolitica produces heat stable enterotoxin, Y-ST. Most of the cases of yersiniosis are sporadic however, the organism has been associated with outbreak involving buttermilk. The control of Yersinia enterocolitica in done by reducing the levels of contamination in the pre process food chain, destruction by thermal processing and adequate domestic cooking and prevention of recontamination.

Shigella

Shigella is a Gram negative, non-motile enterobacteriaceae member. The four subgroups of Shigella differentiated by a combination of biological and serological characteristics are S.dysenteriae, S.flexneri, S.boydii and S.sonnei. The infection caused by Shigella is termed as Shigellosis. The organism is transmitted to humans both by food and water. The infectious dose of shigella is ranging from 1 to 10^4 cells. The symptoms of Shigellosis is characterized by frequent passing of stools which contains blood and inflammatory cells, abdominal pain and fever. The common foods implicated as vehicles for shigella infection are salads, lettuce, milk cheese and cooked rice. The shigella was isolated in foods lioke fish and tamarind and its percentage contamination varied from 1-4%. Infection of Shigella can be controlled by preventing contamination of salads and other foods eaten without cooking. Control of contamination during handling especially after heating should be followed for prevention of shigella infection.

V.Cholera

V.cholera is a Gram negative, oxidase positive, facultative anaerobic, motile curved bacteria. This bacteria is considered both as food and waterborne pathogen. The important species of vibrio with respect to foods are V.cholerae, V.parahaemolyticus and V.vulnificus. The foods which are associated with vibrio are meat and seafoods and the foods which are implicated in food borne disease caused by Vibrios spp are fish, bread, vegetable curry and sandwiches.

Cholera is caused by strains of V.cholerae serogroup O1 and O139. The serogruop O1 is having two biotypes and three serotypes. The biotypes are classic and El Tor and the serotypes are Inaba, Ogawa and Hikojima. The cholera incubation period varies from 6hrs to 3 days. The symptoms of cholera are mainly passage of rice stool water with rapid loss of body fluids and mineral salts like potassium. The excess fluid loss results in dehydration, hypertension and salt imbalance. V. parahemolyticus is also an important pathogen causes diarrhea, abdominal cramps and nausea.

Clostridium Botulinum

Clostridium botulinum is a Gram positive, rod shaped, spore forming anaerobic bacteria produces a foodborne disease called botulinum. This bacteria usually resides in canned foods. Clostridium botulinum produces toxin called neurotoxin. There are four distinct phenotypic groups of clostridia are recognized. Type I and II strains produces toxin types A, B, E and rarely type F. Type III rarely causes illness in humans. The botulinal toxin ingestion results in muscle paralysis. The early symptoms are mainly nausea, vomiting followed by neurological symptoms such as double vision, fixed dilated pupils, speech impediment, difficulty in swallowing etc.

Clostridium Perfringens

Clostridium perfringens is Gram positive, rod shaped anaerobic bacteria and formerly known as Cl. Welchi. It produces toxins from A-E and are responsible for serious diseases in humans such as gas gangrene and enterotoxaemias. The general symptoms of Clostridium perfringens type A food poisoning occurs usually between 12-12 hrs after eating contaminated food stuff. The symptoms are mainly abdominal pain, flatulence, diarrhea, nausea and vomiting. The infective dose is typically varies from 10^6 to 10^8 cells.

Bacillus Cereus

Bacillus cereus is a Gram positive, rod shaped, facultative aerobic bacteria. This bacteria is usually associated with foods like milk, meat, cooked and uncooked rice and its percentage of contamination varies from 5-50%. This bacteria causes both foodborne infection and intoxication. The bacteria produces one or more number of toxins (emetic and diarrhoeagenic toxin) responsible for food poisoning. The emetic toxin produced in foods, causes vomiting, nausea and rarely diarrhoea. While the diarrhoeagenic toxin produced in the GI tract causes abdominal pain and diarrhea but rarely vomiting and nausea. The infective dose of emetic syndrome is 105 -108 cells/g of food where the toxin is pre-formed in the food, in case of diarrhoeagenic syndrome the effective dose is 105 -107 cells require to infect small intestine.

Virus

Virus usually get into the food through contamination with fecal material from infected persons. The common viruses which causes foodborne illnesses are Norwalk like calciviruses and Hepatitis A virus (HAV).

Hepatitis A and E viruses

Hepatitis virus is considered as both food and waterborne pathogen. Acute hepatitis caused by HAV is self-limited and infrequently causes disease that results in death. The infection begins with mild illness characterized by malaise, low grade fever, nausea and head ache. The severe symptoms include vomiting, diarrhea, and arthritis. These symptoms are accompanied by jaundice and the illness usually lasts for 2-6 weeks. The reported case fatality rate in US is 0.3% and varies from 0.004% in persons 5-14 years old and 2.7% in persons older than 49 years. The incubation period for HAV infection is about 28 days but can range from 2-6 weeks.

The prevention strategies to reduce the risk of HAV infection from consumption of contaminated food include prevention or identification of contamination where food is grown, processed, harvested and prepared, virus disinfection of potentially contaminated foods to render them safe for human consumption, immunization to minimize the general risk of infection.

Hepatitis E virus (HEV) is primarily associated with ingestion of contaminated drinking water. The HEV incubation period varies from 22-60 days. The symptoms of HEV infection are nausea, dark urine, abdominal pain, vomiting, joint pain, rash and diarrhea. Quality of the water supply is very important in prevention or control of HEV infection.

Norwalk Like Virus

Norwalk like viruses is the most common cause of foodborne viral illnesses. The virus transmits to human beings through contaminated food or water. Astroviruses, Calciviruses, small round structured viruses belongs to Narwalk group of viruses. Narwalk group of viruses can occur in vomit as well as feces and hence may spread through aerosol droplets. Shellfish have been responsible for some very large community wide outbreaks. Narwalk and Narwalk group of viruses are the most frequently reported acuses of foodborne viral illnesses in UK.

Rotavirus

It has been identified that rotavirus as the most common diarrheal disease causing virus, requiring treatment in children under 5 years of age. Contamination of rotavirus occurs through the use of fecal wastes for fertilization and or sewage effluents for irrigation of crops, washing of fruits and vegetables in contaminated water, preparation or handling of food through rotavirus contaminated hands. The incubation period of rotavirus is 1-3 days and the symptoms include vomiting, fever and respiratory illness.

Protozoan Parasites

Giardia, Cryptosporidium and cyclospora are the protozoan parasites that parasitise both human and non-human hosts. The parasites enter the human body through contaminated food. G.duodenalis also known as G.intestinalis (lamblia) is one of the species of Giardia causes Giardiasis. The symptoms of giardiasis are frequent diarrhea, sometime frothy, greasy and offensive, stool may float on water. Cryptosporidium causes cryptosporidiosis and it is a common cause of acute self-limiting gastroenteritis. The symptoms include flu like illness, diarrhea, vomiting, malaise, abdominal pain, anorexia, nausea and malabsorption. Symptoms usually start after 3-14 days of post infection. Cyclospora causes cyclosporiasis and is a flu like illness. The symptoms are diarrhea, weight loss, low grade fever, anorexia, nausea, vomiting and abdominal pain. The incubation period is between 2-11 days of infection.

Taeniasis and Cysticercosis

Taenia saginata (beef tapeworm) and T.solium (pork tapeworm) are the two known causes of taeniasis in humans. It is more common in where human and domestic animals live together under poor sanitary conditions and undercooked beef or pork is consumed. Abdominal pain and nausea are most common symptoms of teaniasis. The contamination of the environment by taenia eggs in human feces is usually restricted by sanitation. Cysticercosis is the systemic disease caused by the invasion of tissue by the infective karvae (cystecerci) of the tapeworms. T.soilum also cause neuro cysticercosis. Epilepsy is the most common sign of neurocysticercosis.

Helminth Infections

Trichinella is a parasitic worm causes a foodborne disease called Trichinellosis. The mode of infection is from ingestion of meat harboring the larvae of the parasite. Trichinella is nematode consists of elongated, cylindrical, colorless, fluid filled body. Trichinellosis is transmitted in two epidemiological cycles, the domestic and the sylvatic. Domestic cycle consists of cyclic transmission of

Trichinella from pig to pig, or pig to rat to pig and pig to humans. Sylvatic cycle consists of transmission of Trichinella among individuals and among species of wildlife. The symptoms are usually, diarrhea, vomiting, abdominal pain and nausea. Trichinellosis can be prevented by keeping swine herds free from Trichinella infection. Meat and meat products have to processed by freezing, freeze drying, cooking, irradiation to make food safe with respect to trichinellosis.

Toxigenic Fungi

Aspergillus produces toxins called aflatoxins produced by A. flavus, A. parasiticus, A ochraceoroseus and A. nomius. At the level of acute toxicity, the aflatoxins cause liver damage in animals and eventually lead to death. The four different types of aflatoxins produced by Aspergillus are B1, B2, G1 and G2. Aflatoxin B1 is the most toxic among all other aflatoxins. The commodities usually contaminated by aflatoxins are maize and ground nut. Ochratoxin is another mycotoxin produced by Penicillium and Aspergillus species. Ochratoxin comes in three secondary metabolite forms such as A, B and C. Aspergillus ochraceus is a contaminant of wide range of commodities such as coffee, beer and wine. Ochratoxin A is considered as a potent nephrotoxin. Fumonisin is best studied among fumonisins and is producesd by Fusarium moniliforme. The umonisins are mainly widespread in maize and maize products.

Table: Acute toxicities of some mycotoxins.

Compound	Producing Fungi	LD_{50} (mg kg-1)
Aflatoxin B1	Aspergillus flavus	0.5 (dog)
Sterigmatocystin	Aspergillus versicolor	166 (rat)
Patulin	Penicillium expansum	35 (mouse)
Ochratoxin A	Penicillium verrucosum	20 (rat)
Citrinin	Penicillium citrinum	67 (rat)
T-2 Toxin	Fusarium sporotrichioides	5 (mouse)
Deoxynivalenol	Fusarium graminearum	70 (mouse)
Fumonisin B1	Fusarium moniliforme	1-4 (horse)

Toxic Marine Algae

Toxic marine algae produces toxins in sea foods, these include dinoflagellates which produces ciguatoxins associated with major shellfish poisoning syndromes like paralytic shellfish poisoning, amnesic shellfish poisoning and neurotoxic shellfish poisoning. The four important toxin producing genera of algae belong to dinoflagellates are Alexandrium sp., Gymnodinium sp., Dinophysis sp., and Pronocentrum sp. The concentration of as low as 200 cells/ml may produce toxic shellfish. Both proliferating cells and resting cysts contains toxins and act as sources of shellfish contamination. Outbreaks of paralytic shellfish poisoning have been occurring regularly throughout the world. Time of onset of symptoms varies from 15 mins to a10hrs, but usually occurs within 2hrs of consumption of seafood. The clinical features of the disease are neurological, in severe cases the death may occur within 2-25hrs. The toxins produced by Gymnodinium sp. and Ptychodiscus breve causes Neurotoxic shellfish poisoning. The symptoms usually begin 1-3 hours after toxin ingestion and include paraesthesia in the mouth progressing to the extremities, ataxia, gastrointestinal symptoms. The patient may feel hot to cold temperature reversal phenomenon which affects the patients perceptions of touching hot and cold surfaces. There is no specific treatment for this disease.

Biological Hazards Control

Contamination prevention is the most critical control factor to enhance food safety. It is very important to keep foods at refrigerated temperature to prevent bacteria or fungi growing to hazardous levels and foods should be cooked thoroughly before consumption in order to kill all the pathogens present in it.

References

- What-is-food-safety: foodsafety.com.au, Retrieved 16, June 2020
- Food-contamination, mycotoxins-and-food-safety: intechopen.com, Retrieved 06, Feb 2020
- Foodborne-illness, science: britannica.com, Retrieved 08, July 2020
- Food-preservation-methods-food-poisoning, biology: byjus.com, Retrieved 19, May 2020
- Food-processing: newworldencyclopedia.org, Retrieved 02, April 2020

Chapter 2

Food Quality: Evaluation and Control

Food quality refers to the sum of all the properties and assessable attributes of a food item. Food quality evaluation and food quality control are two of its major aspects. This chapter closely examines these key concepts of food quality to provide an extensive understanding of the subject.

Food quality represents the sum of all properties and assessable attributes of a food item. Usually this is done by the three accepted categories of quality: sensoric value, suitability value and health value. All three deal with assessments, that is, judgements with a subjective component. Food quality is an important food manufacturing requirement, because food consumers are susceptible to any form of contamination that may occur during the manufacturing process. Many consumers also rely on manufacturing and processing standards, particularly to know what ingredients are present, due to dietary, nutritional requirements, or medical conditions. Besides ingredient quality, there are also sanitation requirements. It is important to ensure that the food processing environment is as clean as possible in order to produce the safest possible food for the consumer.

The notion of food quality rests on a complex and multi-dimensional concept which is influenced by a wide range of situational and contextual factors. The characteristics contributing to food quality are depicted in the figure below:

The 24th Food and Agricultural Organisation of the United Nations (FAO) Regional Conference for Europe attempted to extend the general definition of quality, which can be found in standards of the International Standardization Organization (ISO), to food quality:

- Historically, quality has been primarily understood as the absence of a defect, fraud and adulteration.
- More recently, quality rests on expected properties such as organoleptic and nutritional

characteristics or resulting benefits. This introduces the need to take the legitimate expectations of users into account and to request operators to do likewise.

- Finally, quality designates desirable characteristics likely to justify added value; for example, forms of production (organic farming, environmental consideration, and animal welfare), production areas (designation of origin) and their associated traditions.

Dimension of Food Quality

Food quality has an objective dimension, which is the measurable physico-chemical characteristic inherent to a food product, and a subjective dimension framed by consumer expectations, perceptions, and acceptance ('fitness for consumption').

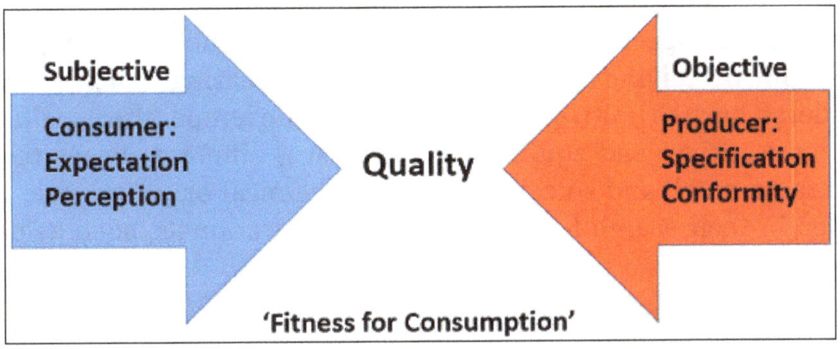

Consumer expectations are based on quality cues (information stimuli) that are either intrinsic (e.g. inferring the ripeness of fruit from its colour) or extrinsic to the product (e.g. advertising, brand image). The better the match between the expectations consumers have before purchasing (expected quality) and the experience they obtain in using the products (experienced quality), the higher the level of consumer satisfaction. It has been demonstrated that positive experience with a brand reduces the risk of dissatisfaction and is frequently used to infer quality of the product*. Consequently, consumers often rely on well-known brand names as quality indicators*. Research has shown that consumers regard the taste and other sensory characteristics of food, health, convenience and the production process (e.g. organic, animal welfare, etc.) as the most important determinants of food quality.

Food Quality Evaluation

Evaluation of foods involves instrumentation and use of physical and chemical techniques to evaluate food quality. Food quality has both subjective and non-subjective aspects. Appearance, texture and flavour are largely subjective attributes, whereas nutritional and bacterial quality is not. Major methods of food evaluation include:

Sensory Evaluation

Foods have several characteristics that require evaluation by sensory methods. The various food attributes that are judged on the sensory scale are flavour, texture, aroma and appearance. Every time the food is eaten, the quality is evaluated or the judgment is made. The reaction of the consumer on the basis of the sensory perception serves as an endorsement or proof of the product acceptance. The sensory evaluation of food is carried out on a scientific basis to ascertain the product formulations or processing techniques that are anticipated to be successful in the market place.

Role of Primary Senses in Sensory Evaluation

The senses of smell and taste have always been used for our guidance for selection of food and beverages. The ability to discriminate between desirable and undesirable foods is apparently as old as human race.

The Human Senses

Psychologists recognise 22 special senses (or subdivisions) in human beings. Humans possess five primary senses for perceiving stimuli. They are sight, hearing, touch, smell and taste. The last two are considered most primitive. Other senses include temperature, pain, visceral hunger, thirst and fatigue, sex (drive) and equilibrium. At least 3 different senses respond to specific chemical stimuli: taste, smell and so called common chemical or pain sense. Humans are primarily sight guided in their search for food whereas other animals, such as dogs and pigs are scent guided.

Sensory Receptors

We experience our environment and many events occurring in our body through specialised sense organs or sensory receptors. These are eye, ear and skin, each as an organ of touch or pressures, the tongue as organ of taste and nose as the organ of smell. Each of these organs responds to a particular range of stimuli and transmits the information to the brain via the central nervous system. Up to a point the response of the sensory cells is proportional to the stimulus intensity.

Physiology of Smell and Taste

Chemical sense receptors for humans are primarily in the nose and mouth, but some receptors for pain (chemical) occur in other locations of the body. Receptor sites for these senses respond to chemical stimuli or the arrival of chemical particles. This is in direct contrast to the physical senses of sight, hearing, touch and temperature which respond to physical stimuli. The sense of smell is substantially more refined than the sense of taste. A typical person requires relatively concentrated solution in order to perceive taste sensation. By contrast, an odoriferous substance e.g. mercaptan may be diluted to the extent of 0.43×10^{-15} mg/Litre of air and still is recognised as such by the sense of smell. The human sense of smell is by far more sensitive than any other chemical procedure or equipment devised so far.

Odour Receptors

Humans are able to sorts out thousands of different kinds of odours and distinguish one flavour from other and even detects slight off flavours. The olfactory receptors are located in the roof of the nasal cavity and are lined with olfactory epithelium which is yellowish in colour with pink respiratory epithelial cells around it. The surface of the olfactory epithelium is coated with a layer of mucous. Embedded in the mucous layer is a mat of fine hair like appendages of the olfactory cells called cilia.

Odour Perception

In order to perceive the odour of a chemical substance the chemical must be volatile. The volatility of a substance depends on its molecular weight and molecular bonding properties. The upper limit of smellability is usually a molecular weight of about 300, although a compound of MW 394 has odour. This suggests that heavier MW substances such as proteins, starches, fats and many sugars are too heavy to be airborne under most circumstances. While molecular bonds vary with chemical compounds and temperature, more volatile molecules are derived from liquids than from solids but volatility of both states matter increases with the increase in temperature. Odoriferous substances must be absorbed or adhered to the chemoreceptor sites in the nose. To be absorbed, a volatile substance must be soluble to some extent in the receptor cell membrane which consists of lipids, proteins and water. Most volatile organic compounds are soluble to some extent in the membrane constituents.

Taste Receptors

Taste receptors are groups of sensory epithelial cells enervated by sensory nerve endings. The taste receptors look like flower buds hence are termed taste buds. Human taste buds are located on moist surfaces within the oral cavity and pharynx. The tongue serves as the major organ of taste. The taste bud sites are in the epithelium of the tip, lateral and dorsal surfaces where they are generally associated with certain papillae. A few non papillae associated taste buds may be found in the soft palate, pharynx and larynx. Fungiform papillae are on the anterior tongue segment and foliate and vallate papillae on the posterior part. These taste buds are embedded within the epithelium of the mucous membrane. The stimulation of taste buds is derived from the sensory component of 3 different cranial nerves: facial, glossopharyngeal and vagus. The facial nerve enervates the taste buds in the anterior of tongue and soft palate. The glossopharyngeal and vagus nerves accommodate sensory enervation of all the remaining taste buds.

Taste Perception

The taste buds, with which any substances (in liquid form) must make contact before a taste sensation occurs, are located in many of these papillae at the tip, sides and the base of the tongue. They differ somewhat in their response to stimuli. The sour taste may be noted chiefly of the tongue along the sides, saltiness along the sides and tip, sweetness generally at the tip, and bitter taste at the base of the tongue. Umami taste buds are present on the tongue and other regions of the mouth. The umami substances are L-glutamate, inosine 5′- monophosphate (IMP) and guanosine 5′-monophosphate.

The Concept of Flavour

Odour and taste combined with mouth feel result in overall concept of sensation referred to as flavour. Flavour has been defined as the sum total of sensory impressions (sensations) perceived when a food or beverage is placed into the mouth. Flavour is the voice of foods and beverages and is composed of taste and odour. If nose or breath is held the flavour cannot be perceived. The odoriferous substances must be sniffed or whiffed rather slowly but strongly while respiration is slowed. During exhalation there is no appreciable smell sensation. Smell sensation may also be perceived by diffusion, however, this is very slow and in any case when breath is held no odour sensation is noted. During mastication molecules of aromatic substances pass to the olfactory area from the mouth. It is believed that during tasting, gentle exhalation through the nose is beneficial because it serves to force volatile molecules towards the olfactory epithelium. The first requirement for tasting a substance is water solubility of the item. The two most primitive tastes i.e. salt and sour show good correlation with molecular structure. The slower response sweet and bitter tastes are less related to chemical formulation. Common salt NaCl and other similar compounds taste salty. However, as the constituents atoms of salts become larger (higher atomic weight) a bitter taste tends to accompany the initial salty taste. Sweetness is characteristics of sugars and is related to the possession of hydroxyl (OH) groups. Bitterness is not easily associated with any structural property. Truly many different flavours are sensed instead of tasted e.g. the true tastes may be sensed with the nose obstructed. In fact when a person has cold, the taste reactions, aside from the particular feel of the food in the mouth are sometimes the only components of flavour detected by the individual.

Classification of Taste and Odours

Strictly speaking we experience only 4 true tastes: bitter, sweet, sour and salty. However, there may be several other taste reactions: alkaline, metallic, watery and/ or meaty. These may be tactual sensations and may not be true tastes. Common experience indicates that the number of perceived odours is many and varied. Several investigators have tried to categorise the many odours. Henning in 1916 categorised them in six groups. Crocker and Henderson in 1927 simplified the group of six into four, arranged in order of preference by most subjects as shown below:

Henning		Crocker and Henderson
Spicy	Cloves, onions, fennel, anise	1. Fragrant or Sweet
Flowery	Heliotrope, geranium	2. Acid or sour
Fruity	Oil of orange, apple, citronella	3. Burnt
Resinous	Balsamic, turpentine, eucalyptus oil	4. Caprylic or goaty
Burnt	Pyridine, tar	
Foul	Hydrogen sulphide, carbon disulphide	

Sight

Some desirable qualities and some defects of food products can be ascertained by careful visual observation. Factors that can be evaluated by the sight are style, neatness and cleanliness of package exterior, attractiveness of product finish, package closures, body and texture, colour and overall appearance, meltdown characteristics (ice-cream and other items) that determine quality. Colour and appearance aspect of food products should not be overlooked because these features may render the product acceptable/unacceptable.

Touch

Touch or mouth-feel plays an important role in texture evaluation. The tongue and palate evaluate feeling of mealiness or grassy butter or sandy defect in ice cream where pressure between the teeth determines the presence of crystallised lactose. The briny defect in butter due to un-dissolved salt or mealy/ grainy in Cheddar cheese are also perceived by the sense of touch. The fingers and thumb also help substantiate the findings of the tongue, roof and other surfaces of the mouth.

Requirements for Sensory Evaluation

A successful implementation of sensory evaluation programme requires three major components i.e. proper laboratory facilities, sensory panels and rigorous training programmes.

Laboratory Set up and Equipments

The sensory laboratory set up normally consists of a reception cum briefing room, panel booths and preparation rooms. Sensory evaluation should be conducted in quite and well lit rooms free from any odours. The dominant motive of constructional details should be to have comfort for concentrated prolonged testing and ease of cleaning. Pleasing neutral shades and maintenance of comfortable temperature and humidity conditions of the whole area or at least the panel room are desirable. The testing area where booths are located should be separated from sample preparation room and wash room and store by a complete partition.

Reception and Briefing Room

It should be so designed as to ensure maintenance of pleasant attitudes and minimise traffic to the booths. Panel members shall assemble here, register, receive the evaluation card and briefed about the test.

Panel Booths

These booths should be located between or adjacent to the reception and preparation rooms, and should consist of test booths or identical design. Each booth may be 75 to 80 cm wide having adequate space to keep samples, drinking water and writing space. It shall be separated by partitions to screen one person from the view of other when they are seated. Revolving stool with back support or chairs should be provided with drinking water, cleansing towels and glasses and basins for convenience. The lighting of booths shall be uniform and glare free and arrangement should be made to provide white or coloured light. A temperature of 20 °C and RH of 62 % in the testing room is ideal. The entry and exit to the panel booth area by independent doors may be useful to avoid any communication between panel members.

Preparation Room

It shall be suitably separated from the testing room and should be equipped for preparing and serving food samples. The room should have the facility for cooking range, hot and cold storage cabinets. The ventilation should be proper and the cooking odours should not penetrate the panel booth area. The samples shall be passed to the test booths through hatch in the partition. The hatch on the service counter should preferably be constructed in such a manner that there shall be no recognition of individual or either side of partition. The laboratory should be flexible enough to handle current and future testing activities, as well as to provide a workable environment for the staff. The use of computers has been recommended for sensory evaluation work. In that case, sensory evaluation laboratory should include space for data processing equipment.

Sampling Requirements

- Sampling should be carried out by a trained and experienced person and it is essential that the sample should be representative of the lot.
- A procedure of sample preparation which is most likely to bring out the difference in the particular quality attribute under evaluation shall be selected. Care should be taken that no loss of flavour occurs and no foreign tastes or odours are imported by the procedure during preparation, storage, serving etc. Foods like hot sauce, spices, vinegar etc. may require dilution with some medium because of their intense physiological efforts.
- The panellist should be allowed to have sufficient sample necessary to make judgement. Unless, only one sample is to be tested, full normal serving quantities shall not be served even though the material is available.
- The temperature of serving should be close to that recommended for the food product. The samples shall be served in utensils of the same type and appropriate size, shape and colour and they shall not import any taste or odour to the sample. The test should be carried out at least one hour before or after lunch.
- Use of materials which are likely to vitiate results, such as smoking, chewing pan (betel vine) and taking intoxicants by a panellists should have a time lapse of at least half an hour before the test. Use of strong odoriferous substances such as cosmetics, flowers and hair oil should be avoided.
- The number of samples served in any one session shall depend upon the nature of the test product and upon the evaluation method used. Foods differ from each other in intensity or quality levels and these may influence the conduct of tests. If interest can be maintained, panel members can evaluate rather large numbers of samples per session particularly with bland or mild flavoured food. Normally the number of samples should be restricted to 6.
- Since coding is necessary to obscure the identity of the sample, a multiple digit code generated from a table of random numbers should be used. Avoid constant use of certain codes or a set of codes to expedite tabulation of results.
- The evaluation card should be clearly printed and the matter should be arranged in logical sequence for examination which is expected under each test. Appropriate terminology without ambiguity shall be used. The evaluation card should be simple, brief, and easy to follow and record what is exactly required. Due weightage should be given to all the sensory attributes.

Selection and Training of Sensory Taste Panel

Panel Selection

By selecting the most stable and sensitive members and training them, one might expect to obtain a small but efficient panel. Selection is important since individuals differ in sensitivity, interest, motivation and ability to judge differences. Discriminating skills may not be general as a good wine taster may not be a good judge of chocolates. Seldom is a judge equally proficient in tasting all qualities and all flavours of foods. Judges are selected based upon following characteristics.

Screening

Some types of screening process are required for selecting panel members including specific tests based on:

- Discriminating differences between solutions or substances of known chemical composition.
- Ability to recognise flavours and odours.
- Performance in comparison with other panel members.
- Ability to discriminate different samples to be used later in the test.

A general approach in screening may be:

- Use as test materials the same products which will be tested later.
- Prepare samples with variations that are likely to be encountered later.
- Adjust the deficiencies so that most will be able to discriminate but some will fail.
- Use score card similar to the ones that will be used later.
- Start with a large number of candidates, with simple selection test.
- Screen on the basis of relative achievement continuing until top ranking group of the desired size is reliably selected.
- Reject at each stage those that are not reliable but select more than actually needed.

A person with previous experience might utilise some of the skills he has developed. He may be able to note and detect differences which are not detected by inexperienced judges. He can often describe the sensory impressions more fully and usually has a better understanding of the terminology employed.

Sensitivity Tests

For general panel selection the candidates can be eliminated on the basis of lack of sensitivity to the senses, attributes, because of poor memory, slow recovery from stimulation and failure to understand the test. Sensitivity to taste or odour appears to be only one of the factors influencing discrimination. In most cases, elaborate tests based on acuity seem unnecessary.

Factors Influencing Sensitivity

Important factors in successful judging are interest, motivation, knowledge and comparison of results, adjustment to the test situation, memory etc. the panel members should be given as much information as possible on the purpose and need of investigation. However, when this information might influence his judgement it should be withheld. A rewards system for maintaining interest is frequently recommended. This may take the form of special pay, time off, special privileges, providing refreshments after panel sessions etc.

Panel Size

The number of judges needed will vary according to the variability of individuals and of the product. For different testing with trained lab judges the panel should consist of 10 to 20 persons with at least 3 or 4 replications per judge per treatment. In order to achieve greater control of the panel, it is possible that 10 judges are better than 20. Expert panels of 5 to 10 are satisfactory. Consumer types tests require more than 80. Institute of food technologists (USA) recommends 3 to 10 for trained; 8-25 for semi-trained and 80 for untrained.

Training

Training should be distinguished from experience. Training steps may be taken deliberately to increase the effectiveness and rate at which the individual assimilates new knowledge. Training is directed towards getting panel members to disregard their personal preferences. It might also be directed to secure recognition of small differences. One of the important aims of training is to obtain homogeneity of response. Sensitivity to basic tastes and odours increases due to training. Training helps judges learn to compare flavours and flavour strengths in spite of time lag between samples.

Environment

Control of environment factors and samples is universally recognised for sensory evaluation of foods. Interruptions and distractions should be avoided during testing. Regularity, quietness, comfortable surroundings, orderliness and smoothness of presentation of samples are important. Judges should record their results independently and should not be able to communicate their impressions to those entering the lab later either verbally or through facial expressions.

Time of the Day

Tests should be arranged when the judges feel their best. Generally a time of 10:15 to 11:45 AM and 3:00 PM have been recommended. At these times the influence of breakfast has passed but the subject is not hungry yet. No significance differences have been observed when the tests were conducted at 11 am or 3 pm.

Masking

Masking refers to intentional minimizing of colours, taste or odour properties so that the differences between samples can be evaluated with less interference from the variable which has been minimised i.e. use of coloured containers, coloured lights etc.

Preparation for Sensory Analysis

It is pertinent to discuss proper preparation of food, cooking procedures, methods of detecting, flavour pick up and the standards used. The normal procedure is to test the food under conditions approximately the same as found under normal consumption i.e. bread should be dry, butter solid, vegetables whole etc. within a set of samples, the standard should also be presented as an unknown. The samples presented must be exactly alike. Among the factors to be considered are visual appearance, sample size, temperature, utensils, pouring, coding, order, instructions and rinsing. The ability to taste may be affected by substances prior to flavour evaluation. Samples should be the same in form, consistency, colour and appearance. Sample size should be sufficient for judges to taste with confidence. Sufficient sample to give a feeling of mouth-fulness is recommended. Uniformity of temperature is very important. All samples should be served in the containers of the same size, shape and colour. The utensils should not impart a taste or odour to the samples. Many detergents leave an odour (since they are perfumed) unless is thoroughly rinsed. Samples should be placed in the container in a uniform/ aesthetic manner. All samples presented should be coded to avoid giving information to the panel. Three digit random numbers are ideal. While some investigators have found no effect from order of serving but usually the first sample is assumed to be standard against which the rest of the samples are rated. Ideally, strong flavoured samples should be served last as they frequently over sensitize the taste buds and recovery is not quick. It may not be necessary to give instructions to a trained panel but this is not true for all panels. Questions arise as to method of swirling the glass, chewing, smelling, to swallow or not, how many samples to warm up, to rinse between tasting or not and the time allowed for each of these. The best practice seems to be allow the judges to use his preferred method.

Sensory Tests

Sensory tests are conducted to meet the following purpose:

- Select qualified judges and study human perception of attributes.
- Correlate sensory with chemical and physical measurement.
- Study processing effects, maintain quality, evaluate raw material selection, establish storage stability or reduce costs.
- Evaluate quality or determine consumer reaction. Each of these purposes requires appropriate tests. There are a substantial number of test methods and new methods continue to be developed. These are classified into following three broad categories:

Category	Test Type
Discriminative	Difference: Paired combination, duo trio, triangle
Descriptive	Descriptive analysis: Flavour and texture profile, QDA
Affective	Acceptance preference: 9 point hedonic

Discriminative Testing

This is the one of the most useful analytical tools available to the sensory professionals. It is on the basis of a perceived difference between two products that one can justify proceeding to a descriptive test. Within this general class are a variety of specific methods e.g:

- Paired comparison test.

- Duo trio test.
- Triangle test.
- Multiple sample test.
- Other test methods such as dual and multiple standard tests.

The main objective of all these methods is to answer a simple question. "Are the products perceived as different"? Obviously the response to this question can have major consequences. If the conclusions from a discrimination test are to be accepted by management as reliable, valid and believed, then it is important that each test be conducted with proper consideration for all aspects of the test design, product preparation and handling, implementation, data analysis and interpretation.

Paired Comparison Test

The paired comparison procedure is the earliest example of the application of discrimination testing to food and beverages evaluation. It has also been used successfully for determination of thresholds for basic taste solutions. The paired comparison test is a two product test and the panellist task is to indicate the one that has more of a designated characteristic such as sweetness, tenderness or skinniness. This method is also identified as a directional paired comparison test. The paired comparison test is relatively easy to organize and to implement. The two coded products (AA, BB) are served simultaneously and the subject has to decide whether there is difference or not. Another version of the paired test is the A-not-A procedure. The subject is presented with a single sample for evaluation, which is then replaced by a second sample. The subject then makes a decision as to whether the products are the same or different. This particular test procedure has considerable merit in those situations where non test variables such as a colour difference may influence results.

Duo Trio and Triangle Test

The duo trio is modified paired comparison test in which one sample is presented first, followed by two coded samples one of which is identical with the standard. The judge has to identify which of the two is identical to the standard. This procedure is used in quality control and for selection of judges of superior discrimination. In triangle test, two identical and one different sample are presented simultaneously and the judges are asked to indicate the odd sample. The chance probability associated with the triangle test is only one third, which accounts for its claim of greater sensitivity. The triangle test is a more difficult test because the subject must recall the sensory characteristics of two products before evaluating a third and then making a decision. Triangle test is also used to select judges.

Multiple Sample Test

Test involving more than three stimuli are classified as multiple sample tests. They may have equal or unequal numbers of each stimulus. When they are applied as true difference tests, the judge is required to separate the sample into two like samples. When they are applied as directional tests, the judge is asked to identify the groups of given criterion. Difference test designs involving more than three stimuli have had only limited use. The limitation is based on the increase in psychological complexity and physiological fatigue which accompanies an increase in the number of stimuli. In addition, large quantities of samples are required and more time is needed for the observer to make a decision. These tests appear to be most applicable to visual discrimination, where the judge does not rely on memory and fatigue is almost non-existent.

Dual Standard Test

The dual standard method was proposed for use in quality control situations. The subject is served four products. Two are identical as reference A and B and two are coded. The subject must match the reference product with the coded product. The designation of the two references could reflect whether product is in quality control limits or is outside the limit.

Descriptive Analysis

A descriptive test involves relatively few judges who have been screened according to the category of product. Training is primarily focused on development of descriptive language which is used as a basis for scoring the product. Other important activities which are part of training include, the grouping of attributes (i.e., appearance attributes, aroma attributes and so on), developing a definition for each attribute, identifying helpful reference for use during training and familiarizing the subjects with scoring procedure. There are numerous applications for descriptive analysis including, monitoring competition, storage stability, product development, quality control, physical/ chemical and sensory correlation etc. Depending upon the test methods used the training can be quite different.

Flavour Profile

The flavour profile method is the only formal qualitative descriptive procedure and is probably the most well-known of sensory test methods. This method utilizes a panel of four to six screened and selected subjects who first examine and then discuss and the product in an open session. Once agreement is reached on the description of the product the panel leader summarises the result in report form. The method has considerable appeal because results could be obtained rapidly and would obviate the need for statistics. Only repeated analysis (10, 15 or 20 times) with the same test product will provide uniform results by a whole panel. The profile method concentrates on the entire flavour of a product and the individual attributes of a flavour in relation to each other. The dimensions of flavour analysis by the profile method include:

- Perceptible aroma, taste, flavour and feeling factors (called 'character notes').
- Degree of intensity of each factor, graded on the following scale:

O	Not present
X	Just recognizable or threshold
1 or +	Slight
2 or ++	Moderate
3 or +++	Strong

- Order in which these factors are perceived.

- After taste.
- Amplitude of flavour, graded on following scale:

X	Very low
1	Low
2	Medium
3	High

Texture Profile

Texture profile is defined as "the sensory analysis of the texture complex of a food in terms of its mechanical, geometrical, fat and moisture characteristics, the degree of each present and the order in which they appear from first bite through complete mastication. Each texture term should be well defined. For each parameter, a scale of 5, 7, 8 or 9 categories and for each category (of a scale) there is a specific product that represent the particular sensation. This is a descriptive method. This includes description analysis from a structural point of view. Structure is referred as the development of the descriptive terminology, the scales for recording intensities, and the word/ product anchors for each scale category. Once this structure is completed, the product evaluation procedure follows the flavour profile approach with panelists discussing results in order to reach a conclusion.

Quantitative Descriptive Analysis

The quantitative descriptive analysis (QDA) method was developed with an approach that was primarily behavioural in orientation with a consensus approach to language development, use of replication for assessing subject and attribute sensitivity. The development of method evolved from a number of considerations to ensure that it would:

- Be responsive to all the sensory properties of a product.
- Rely on a limited number of judges for each test.
- Use judges who are qualified before participation.
- Be able to evaluate multiple products in individual booths.
- Use a language development process free from leader influence.
- Be quantitative and use repeated trials design.
- Have a useful data analysis system.

In a QDA test, the judges evaluate all of the products on an attribute on more than a single occasion. Descriptive Analysis is used for storage testing, product development, quality control an instrument sensory relationship.

Affective Testing

Acceptance preference a valuable and necessary component of every sensory programme is performed at consumer's levels. It refers to measuring liking or preference for a product. Preference can be measured directly by comparison of two or more products with each other. Indirect measurement of preference is achieved by determining which product is scored significantly higher than another product in a multiproduct test, or which product is scored higher than another by significantly more people. The two methods most frequently used to measure preference and acceptance are the paired comparison test and the nine point hedonic scale. Other methods are either modifications of these two methods or are types of quality scales for example excellent to poor and palatable to unpalatable.

Hedonic Scale

The nine point hedonic scale has been used extensively since its development with a wide variety of products and with considerable success. The scale is easily understood by consumers with minimum instruction and a product differences are reproducible with different groups of subjects. The results from use of this scale are most informative since computations will yield means, variance and frequency distributions, by subject. An example of scale is given below:

Like extremely	9
Like very much	8
Like moderately	7
Like slightly	6
Neither liked nor disliked	5
Dislike slightly	4
Dislike moderately	3
Dislike very much	2
Dislike extremely	1

Scoring

The most frequently used of all sensory testing systems is scoring because of its diversity, apparent simplicity and ease of statistical analysis. Scoring methods have been most extensively used for the product development and improvements. Score cards based on 100 points are generally used for judging and grading of food products. It believed that numerical rating tests give more complete information that either ranking tests or descriptive rating tests. But the judges must be trained. The score card must be properly developed giving due weightage to all the sensory attributes.

Objective Evaluation

Objective evaluation supplements or reinforce the data obtained subjectively through sensory evaluation.

Basic Guidelines:

- Conduct all objective tests appropriate to the experiment for which equipments is available; For example, checking the pH of mixtures prior to and after heating. This information often is valuable in explaining results. Flow properties of batters or certain other mixtures prior to heat treatment can be measured and that information used in interpreting and explaining results.

- Obtain necessary testing devices: In the preliminary testing phase of an experiment, analyse all steps in the preparation of the product and study the final product to determine whether there are specific characteristics that might be tested objectively if additional testing equipment could be procured of developed for the experiment.
- Be meticulous about maintenance of objective equipment: Before using any equipment, the good researcher will check to be certain the machine is operating correctly in all aspects.
- Carefully define the samples to be used for objective testing: A template of the item being tested often is an essential tool in obtaining comparable samples for objective tests. The dimensions of pastry or cookie samples being tested for tenderness on the shortometer must be identical. To obtain this sample, the thickness of the mixture prior to baking must be controlled precisely.
- Establish operating conditions for objective testing: For example, the temperature of a starch paste being utilized for a linespread test must be specific and controlled so that the effect of temperature on viscosity of starch pastes is not an uncontrolled variable in the measurement. Samples to be controlled in size, storage and temperature. Each experiment needs to be designated to eliminate uncontrolled variable in objective testing.

Tests Used for Objective Evaluation

Chemical Methods

Chemicals are estimated in food spoilage like peroxides in fats. Adulterants in food e.g., presence of starch in milk, metanil yellow in turmeric powder and loss of nutrients during cooking can be estimated.

Physico-chemical Methods

- Measurement of hydrogen ion concentration can be found by the use of pH meter. It utilizes a glass indicating electrode and a reference electrode to complete the electrical circuit.
- Digital Salt Meter: This refractometer has electrodes built in the measurement unit and it indicates salinity percentage in digits in 3 seconds with 1 ml sample dripped on it.

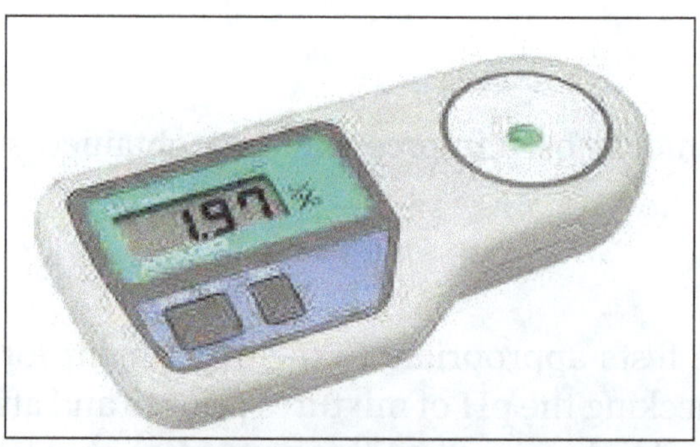

Digital salt meter

- Sugar concentration can be found by refractometer. It is used to determine the concentration of a sugar solution. Light is refracted as it passes through sugar solution, with the specific values being calibrated in degrees, Brix, an indication of the percent of sucrose in the solution.

Brix or Balling hydrometer gives directly the percentage of sugar by weight in the syrup. It is always necessary to make a temperature correction since the hydrometers are usually calibrated at 20'C. each instrument used by canners usually covers a range of only 10' Brix, e.g., 10-20, 20-30, 30-40, 40-50, 50-60 Brix respectively and are graduated in 1/10th divisions. Brix is defined as percent sucrose measured by a Brix hydrometer.

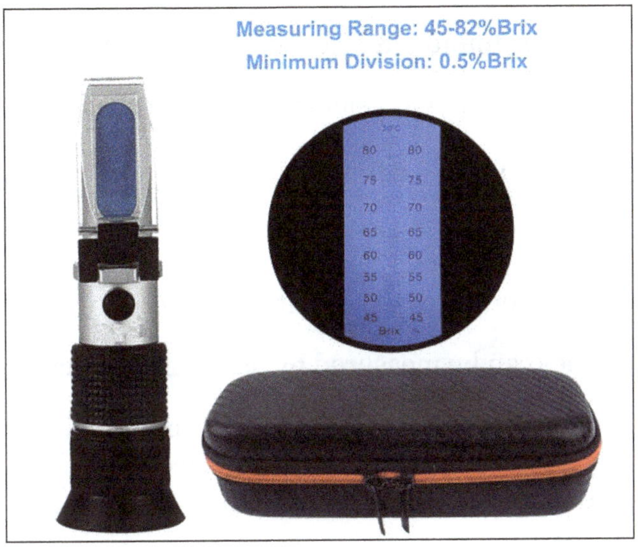

Products Used at Different Brix

Measurement range in Percentage	Main samples to be Measured
Brix 0.0 to 32.0	Fruit juice, softdrinks, lactic drinks
Brix 28.0 to 62.0	Ketchup, Soy sauce, Sauce
Brix 58.0 to 90.0	Liquid sugar, honey
Brix 45.0 to 82.0	Condensed milk, Concentrated fruit juice, jam
Brix 0.0 to 10.0	Puree, tomato juice
Brix 0.0 to 20.0	Fruit juice, soup, word
Brix 0.0 to 50.0	Ketchup

Since continued use of hydrometers in hot syrups affects their accuracy, they should be checked frequently by more accurate instruments.

- Polariscope is used for quantitative analysis of sugar.
- Butyrometer: it is an instrument consisting of a calibrated glass tube for measuring the butter content of milk. The milk is mixed with a certain volume of ether which dissolves the butter. Then an equal volume of alcohol is added. The butter floats on the surface in

the form of an oily layer and its thickness, measured by the graduation of the tube, clearly shows the proportion of butter.

Microscopic Examination

Some properties of foods depend on their structure and valuable information can be obtained by microscopic examination. Some examples are given below:

- Type of organisms present in fermented products like idli batter.
- Examination of starch cells under the microscope for identification.
- Spoilage of the food can be found out by observing the organisms under the microscope.
- Size of crystals in sugar is related to smoothness of the product.
- Number and size of the air cells in batters and foams.

Physical Methods

- Weight: Weight of a food indicates the quality like in case of apple or egg.
- Volume: Liquid volumes can be measured by using measuring cups.

  Solid food volume can be found by displacement method. In this method, the volume can be calculated by subtracting the volume of seeds held by a container with a baked product from that of volume of seeds without the baked product. Usually mustard seeds are used.

- Specific Volume: The determination of specific volume of any product should be done with care and average of replicates are to be taken since experimental errors are likely to be large. Measurement of bulk volume in a porous and spongy product like idli is difficult. The volume may be measured by displacement with solvents like kerosene. The idli is given a momentary dip in molten wax to seal off the pores. Increase in volume is taken as the measure of its bulk volume.

  $$\text{Specific Volume} = \frac{\text{Bulk Volume}}{\text{Wt. of the substance}}$$

- Index of volume: It can be found by measuring the area of a slice of food with a planimeter. It is important to use a slice that is representative of the product such as a centre slice.

  Index to volume is a measurement made by first tracing detailed outline of a cross section of the food. This tracing can be done with a sharply-pointed pencil or a pen or by making a clear ink blot of the cross section. The ink blot is made simply by pressing the cross section of the sample lightly onto an inked stamp pad and then making the imprint of the inked sample on paper. A planimeter can also be used to trace the entire outline of the sample, being careful to follow all indentations and protrusions so that the final measure recorded on the planimeter represents the circumference of the slice.

- Specific gravity: It is a measure of the relative density of a substance in relation to that of water. The measurement is obtained by weighing a given volume of the sample and then dividing that weight by the same volume of water. This technique is used for comparing the lightness of products physically unsuited to the volume measurements e.g., egg with foams. Potatoes with low specific gravity (waxy type potatoes) have cooking characteristics different from those of potatoes with a comparatively higher specific gravity.

- Moisture: Press Fluids: Initial weight of the sample is noted. After the appropriate pressure has been applied for a controlled length of time, the sample is again weighed. The difference between the two weights represents the amount of juice contained in the original sample e.g., juiciness of meats, poultry and fish.
- Drying: The weight of the original sample is determined and then the food is dried until the weight remains constant.

$$\text{Moisture Content} = \frac{\text{Initial Weight} - \text{Dried Weight}}{\text{Initial Weight}} \times 100 = \%$$

- Karl Fischer Titration: In 1990, Karl Fischer showed that food to be analysed by this method is homogenized in a high-speed blender at speeds up to 7,500 rpm to release the water and the water is titrated with Karl Fischer reagent until all the water has reacted with the reagent. The calculation for water content is handled by a microprocessor, which is built into the machine. It is costly but gives quick response.

Wettability

Baked products can be tested for moisture level by conducting a test for wettability. For this test, the sample is weighed before being placed for 5 seconds in a dish of water. Immediately at the end of the lapsed time, the sample is removed from the water and weighed again to determine the weight gain. High moisture retention is synonymous with good wettability, a sign that a cake probably will be considered to be appropriately moist when judged subjectively.

Cell Structure

Cell structure of baked products is an important characteristic to measure the uniformity, size and thickness of cell walls. Photocopies of cross-sectional slices give this valuable information. This techniques gives third dimensional view into the cells on the cut surface of the sample and give the actual size clearly.

- Size of the grain: This can be found by using photography or ink prints with stamp pad or sand retention e.g., idli. Retention of sand is more if the grains are coarse. Cut the idli into 2 pieces and take one piece and press it on the stamp pad and take an impression on the paper. Ink prints may be less clear but satisfactory for some purposes.
- Photography: This may be colour or black and white. They may not represent the sample size so a marked ruler should be kept adjacent.

Measurement of Colour

Colour is the first quality attribute a consumer perceives in food. Change of colour is generally accompanied by 8auté8 changes.

- Colour dictionaries: The dictionary of Maerz and Paul is most commonly used. The dictionary consists of 56charts. Seven main groups of hues are presented in order of their spectra. For each group there are 8 plates. In place of colour dictionary, colour reproduced on secondary standards such as painted test panels, rings, discs or plastic models may be used.

  A mask of neutral grey having two opening is used. The size of each opening should be equal to the size of of the individual colour path in the sheet. An opening should be placed over the sample and the other over different patches on the chart until a match is achieved and the colour is noted.

- Disc Colourimeter: Here the discs have radial slits so that a number of them may be slipped together with varying portions of each showing. The discs are spun on a spindle at about 2700 rpm so that the colours merge into a single hue without flickering. The test sample is placed adjacent to the spinning disc under controlled illumination and both are viewed simultaneously.

- Coloured Chips: A simple method is to match the colour of the food with the colour chips or colour glass, chart or colour tiles. This method is not very satisfactory as it is difficult to match the food with one small block of colour or the chart. The data are difficult to tabulate and analyse also.

- Spectrophotometer: Visual matching of colours is subject to shortcomings of human observers. To overcome this spectrophotometer can be used. In this, tub e with the liquid is placed in a slot and light of selected wavelength is passes through the tube. This light will be differentially absorbed depending upon the colour of the liquid and the intensity of the colour. Two liquids of exactly the same colour and intensity will transmit equal fractions of the light directed through them. If one of the liquid is a juice and the other is the juice diluted with water, the latter sample will transmit a greater fraction of the incoming light and this will cause a proportionately greater deflection of the sensing needle on the instrument. Such an instrument can also measure the clarity, cloudiness of a liquid depending on the amount of light the liquid allows to pass.

Instruments Used for Texture Evaluation

Various instruments are used to measure the texture of liquids, semi-solids, Rheology is defined as the science of deformation and flow of matter. It has three aspects – elasticity, Viscous flow and plastic flow.

The science of rheology deals with the measurement of various mechanical properties of foods. A study of rheological properties of foods is important for two reasons:

- To determine the flow properties of liquid food stuffs.
- To ascertain the mechanical behavior of solid foods when consumed and during processing.

Instruments Used for Liquids and Semisolids

The resistance or internal friction to the flow of liquids is normally known as viscosity. Viscosity or consistency is an important factor in influencing the quality of a large number of food products. The more important among these are cream style corn, salad creams, tomato products, jellies, jams, mayonnaise, syrups, and fruit pulps where the acceptability largely depends on their having proper consistency or viscosity. Measurement of this factor for the raw material or the product at various stages of manufacture serves as an aid in checking or predicting the consistency of the final product. Further, such quality control measurements also serve as indicators in calculating the amount of an ingredient (thickening agent, etc) that should be added in a particular food product. Duration and amount of heat applied in a process may also be suitably regulated to some extent by viscosity measurement as heat penetration and consistency are closely interrelated.

Percent Sag

The depth of a sample such as jelly is measured in its container by using a probe. The product then is unmoulded onto a flat plate. The greater the percent sag, the more tender is the gel.

$$\text{Percent sag} = \frac{\text{Depth in container} - \text{Depth in plate}}{\text{Depth in container}} \times 100$$

Baker's Jelmeter

The relative viscosity of pectin extracts or solutions is the ratio of the time of flow of the pectin solution through a given orifice compared to the time of flow of an equal quantity of water at the same temperature. A simple capillary pipette called the jelmeter has been developed by Baker for such viscosity determination.

$$\text{Relative Viscocity at T'} = \frac{\text{Time in seconds for the extract to flow through}}{\text{Time in seconds for the water to flow through.}}$$

The various methods based on different equipment used for measuring jelly strength can be divided into two broad groups. The methods under the first group measure the breaking strength of jellies when they rupture after exceeding their elastic limits, while those of second group measure jelly strength by taking into account the deformation of jellies within their elastic limits.

Stormer Viscometer

It is used to measure the viscosity or consistency of certain food products and to give an index of the resistance of the sample to flow. The number of seconds required for the rotor to make 100 revolutions has been used to measure the consistency of some food samples.

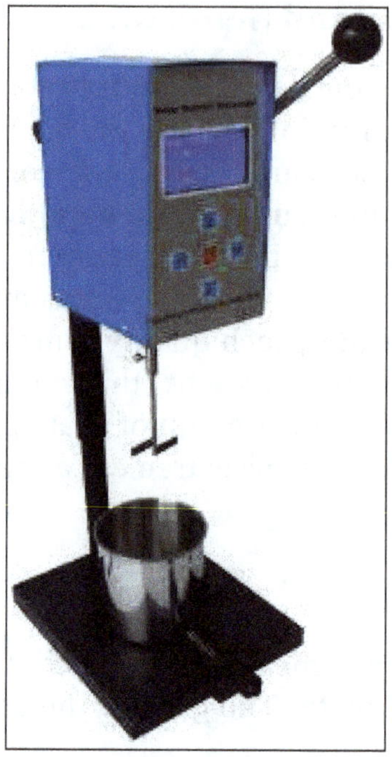

Brookfield Synchrolectric Viscometer

This is based on measurement or resistance to rotation of a spindle immersed in the test material. This can be used successfully in measuring the consistency of custards, pie fillings, tomato products, cream style corn, mayonnaise, salad dressings and dairy products.

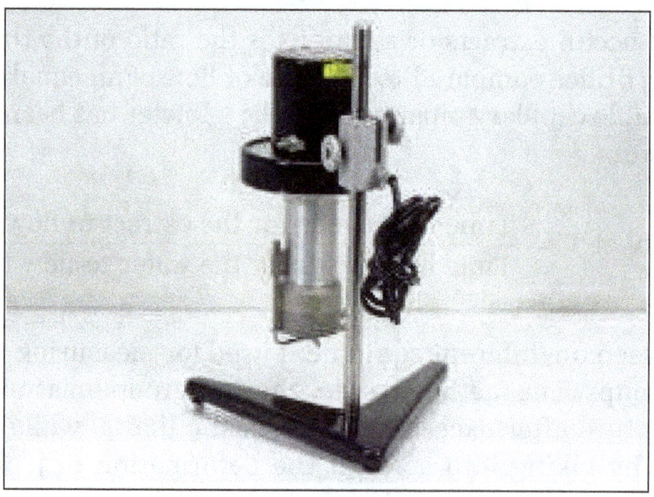

Penetrometer

A penetrometer also may be used to measure tenderness of some foods. This device consists of a plunger equipped with a needle or cone that is allowed to penetrate the sample by gravitational force for a selected period of time. The larger the reading the longer the distance the more tender is the product. Gels and many baked products are particularly well suited to tenderness measurement using the penetrometer.

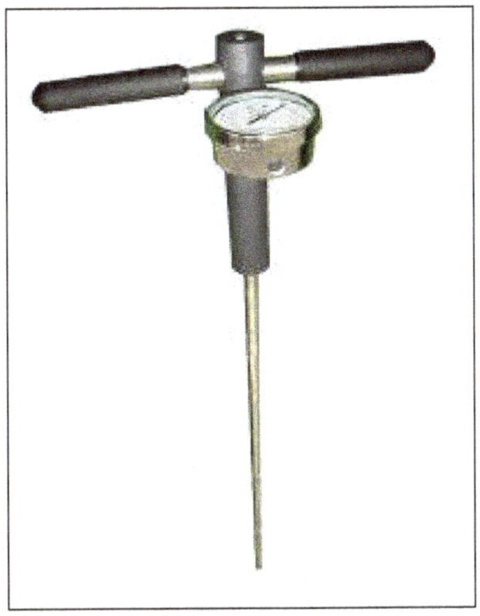

Brabender Farinograph

This is used to measure the plasticity of wheat dough for preparing bread products. It is designed to study the physical properties of the dough by recording the force required to turn the mixer plates through the dough. The force required increases as the solution develops during mixing and later decreases as solution is slowly broken down by over-mixing.

Food Quality Control

Quality control is the maintenance of quality at levels and tolerance limits acceptable to the buyer while minimizing the cost for the vendor. Scientifically, quality control of food refers to the utilization of technological, physical, chemical, microbiological, nutritional and sensory parameters to achieve the wholesome food. These quality factors depend on specific attributes such as sensory properties, based on flavor, color, aroma, taste, texture and quantitative properties namely; percentage of sugar, protein, fibre etc. as well as hidden attributes likes peroxides, free fatty acids, enzyme.

Although, quality attributes are many, not all need to be considered at every point in time for every particular product. It is important to always determine how far relatively a factor is in relation to the total quality of the product. The quality attribute of a particular product is based on the composition of the product, expected deteriorative reactions, packaging used, shelf life required and the type of consumers. The most important element and ultimate goal in food quality control is protecting the consumer. To ensure standardization of these procedures, food laws and regulations cover the related acts affecting the marketing, production, labeling, food additive used, dietary supplements, enforcement of General Manufacturing Practice (GMP), Hazard Analysis and Critical Control Point (HACCP), federal laws and regulations, factory inspections and import/export inspections.

Importance of Food Quality Control (QC)

The most important quality factor of processed food is safety and reliability followed by "deliciousness" and "appropriate price". The colossal loss a food industry will record if defective products were rejected or recalled, as well as the damaging effect on the company's image and public trust justifies the need for food quality control. For this reason, quality assurance should be a corporate goal, and should stem from the uppermost management level to the least staff of the industry. The Plan, Do, Check, Action (PDCA) cycle should be used when quality control is implemented.

The Food Supply Chain

The food supply chain is a net chain that food moves along the peasant household, the processing industry, the distribution centre, the wholesaler, the retailer as well as consumer. In general, the food supply chain is composed of five links: product material supplying link, the production processing link, the packing storage and transporting link, the sale link and the consumer expending link with each link involving related sub-links and the different organization carriers. The supply chain is a networked structure, composed of flow of the physical, the information, the finance, the technical, the standardized, the security and the value-added connections.

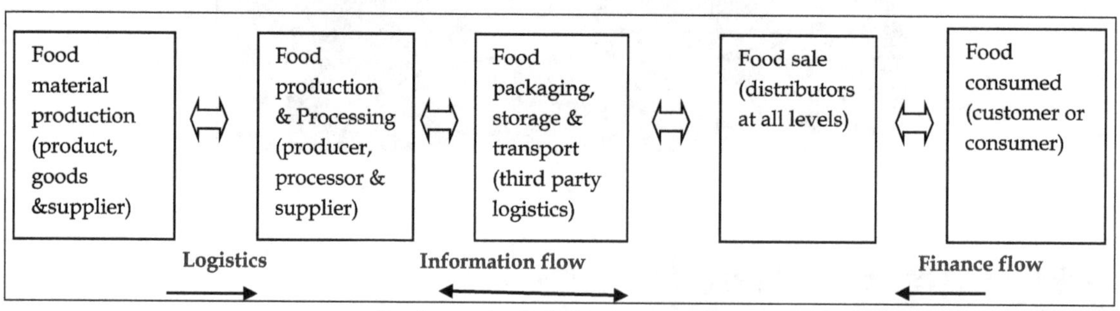

Food supply chain basic structure.

Quality Procedures

Along the supply chain, food is unavoidably exposed to numerous hazards. Therefore, knowing the risk factors at each phase of the supply chain assists in ensuring that an effective and comprehensive quality system is put in place. To guarantee food quality, it is important to ensure that, all the steps of the supply chain are carried out strictly, with care, and according to the standard operating procedures. At each phase of the supply chain, the potential risks, responsibilities, and how they can best be addressed must be fully explored.

Food Contamination

Contamination of food can occur at any of the phases of the food supply chain and these will be expounded under the following broad categories:

- Physical
- Chemical
- Microbiological
- Other contaminants

Physical Contamination

One of the major physical contaminations is adulteration. It is the mixing of inferior quality material with the superior product, thereby reducing the nature, quality and originality in taste, color, odor and nutritional value and ultimately causing ill effects on the health of the consumer. The main motive of adulteration and of course the "adultratee" is to gain undue advantage and most often profits. Almost all the food stuffs being sold in the market are prone to adulteration, but main food products that are often heavily adulterated are spices, milk products, edible oil, beverages drinks, sweets, pulses, sugar, processed foods, rice and cereal products like flour. The table below shows some common foodstuff and their adulterants.

Table: List of common foodstuffs and their adulterants.

Types of Foodstuff	Main adulterants
Milk and Milk products	Water, refined oil, separated (cream less) skim milk, solution, flours, animal fats, starch, dextrin.
Butter and Ice cream	Hydrogenated fats and animal fats, Artificial sweeteners, jelling agents, other fats, water nut flour, non-permitted colours.
Vegetable Oils and Fats	Coffee powder exhausted coffee powder starch, roasted dates, tamarind seeds, tea residues, other leaves with added color.
Dry Beverages	Coffee powder exhausted coffee powder starch, roasted dates, tamarind seeds, tea residues, other leaves with added color.
Wet Beverages	Spurious narcotics and other concentrated liquids, Artificial sweeteners (Saccharin), mineral acids, Colored invert sugar, high concentrated sugar solution, Cheap liquids, wine and beer water and juices.
Spices and Condiments	Whole turmeric Coating with lead chromate or coal tar dye starch or talc colored yellow with coal tar dye, Starch colored brown with coal tar dye, Coriander seed Other green colored seeds Powdered bran or sawdust colored green with dye, Starch colored red with coal tar dye, Argemone seeds, Artificial cumin seed like product, Dried papaya seeds and other plant gums.
Cereals and their products	Wheat Stones, straws, low variety grains,Talc, chalk powder, tapioca flour, Maize flour, low variety rice and semolina.

Pulses and their products	Yellow maize flour and Tapioca flour.
Miscellaneous Items	Seeds or nuts broken and colored, Pickle, jam, chutney and squash of cheap products. Also Sweets Artificial sweeteners and tapioca flour.

Chemical Contamination

Chemicals, which elicit harmful effects when consumed by animals or humans, are said to be toxic. The use of chemicals in the production and processing of food and food products not only affects the quality, but also disguises the deterioration and constitutes deliberate adulteration which is potentially very harmful to the health. It is advised that food additives like colouring matter, preservatives, artificial sweetening agents, antioxidants, emulsifiers/stabilizer, flavors/flavouring enhancers etc., if used should be of approved quality and processed under good manufacturing practices.

Table: List of chemical adulterants in food.

Types of Foodstuff	Main adulterants (Chemical formula and type)
Fish, Meat & Dairy Products, Eggs and Poultry	PCBs (polychlorinated biphenyls)
Citrus and other fruits	Juice Production and flavourings (multiple organic acids)
Oil and fat solvents	Dichloropropane ($CH_3CHClCH_3Cl$)
Alcohol Beverages	Ethanol (CH_2CH_2OH)
In Asian-style sauces such as soy, oyster, mushroom sauces, etc.	3-MCPD (3-monochloropropane-1,2-diol)
Potato tubers	Glycoalkaloids (GA)
Unclean soft wheat for use. in non-staple foods and baby foods	Deoxynivalenol (Vomitoxin)
Fruit brandies and liqueurs, wines, distilled spirits	Ethyl carbamate

Microbiological Contamination

Microbiological contamination of food is perhaps the most prevalent health problem in the contemporary world. To ensure good quality and safe food therefore, microbiological criteria should be established and freedom from pathogenic microorganisms must be ensured, including the raw materials, ingredients and finished products at any stage of production/processing. Accordingly the microbiological examination of the foods products has to be adopted widely. The microbiological criteria must be applied to define the distinction between acceptable and unacceptable foods. Food poisoning often results from the consumption of old, used, residual, fermented or spoiled food, as these may be contaminated with bacteria or other microorganisms, hence toxic. Infants and children are more susceptible to food poisoning and care should always be taken when giving them food. Gastroenteritis is caused by food contaminated with the enterococcus, streptococcus faecalis, which is frequently found in the human intestinal tract. Food poisoning may also be caused by inadequately refrigerated food contaminated with microorganisms such as Clostridium perfringens which grows well in the alimentary canal producing the poisoning within 8-12 hours after the ingestion of contaminated food. Bacillus cereus, a gram positive, aerobic, spore-forming organism has been reported to be the etiologic agent in numerous food poisoning outbreaks. Some of the fatal effects of microbial contamination of food include; liver cancer is which is high in some countries due to aflatoxin. Flavism (hemolytic anemia) is caused by eating broad beans or by inhaling the pollen of its flower. In severe cases death may occur within 24-48 hours of the onset of the attack. Lathyrism is a disease, which paralyses the lower limbs.

Other Contaminants

Metals are one of the many unintentional contaminants of food. When present beyond trace amounts, they are toxic. They find their way into food through air, water, soil, industrial pollution and other routes including food utensils. Common examples include, enamelware of poor quality which contributes antimony and galvanized utensils leaching zinc. A major source of tin contamination is tin plate, which is used for making containers for all types of processed foods. It has been shown that, a small quantity of metal is added when food is cooked in aluminum utensils. Although, copper is an essential trace element required by the human body, but copper-contaminated food is toxic. Fumigants used to sterilize food under conditions in which steam heating is impractical contaminate food. This is because, they may react with food constituents to produce or destroy essential nutrients. For example, ethylene oxide, a commonly used fumigant, reacts with inorganic chloride to form ethylene chloro hydride, which is toxic. Some solvents like trichloro ethylene used for the extraction of oil from oil seeds react with the foodstuff being processed resulting in the formation of toxic products. During processing of food, lipids may undergo numerous changes on prolonged heating, oxidative and polymerization reactions could take place, thereby decreasing the value of the processed products.

Smoking of meat and fish for preservation and flavouring is an old practice. But this processing contaminates the food with polycyclic aromatic hydrocarbons such as benzopyrene, many of which are carcinogenic.

Lubricants, packing materials etc. also contaminate foods. A number of chemicals are intentionally added to foods to improve their nutritional value, maintain freshness, impact desirable properties or aid in processing. They also contaminate food if excessive in quantity.

Methods of Food Quality Control

In addition to ensuring safe and health food for the consumer, product manufacturers and service industries have realized that competition in a global market require a continual and committed effort towards the improvement of product and service quality. Therefore, they follow the process improvement cycle comprising PLAN (plan improvement), DO (implement plan for improvement), CHECK (analyze collected data) and ACT (take action). Quality control process consists of raw materials, in-process, product and service. The major factors in process that cause variability in quality of finished product are people, equipment and methods or technologies employed in the process. Use of proper statistical process control methods is also vital for assurance of the product quality. Usually, the value of quality characteristics is used to provide feedback on how processes may be improved. Statistical quality control comprises the following procedure:

- Finished product is measured.
- Sampling occurs for days or weeks.
- Lot is either accepted or rejected based on information from sample.

Contrary to statistical quality control, statistical process control methods focus on identifying factors in process that cause variability in finished product, eliminating the effect of these factors before worse product is manufactured, and control charts give on-line feedback of information about process. Food quality control measures have continuously improved since the 20th century, owing largely to the implementation of good practices, quality systems and increased traceability in food production. Ever since microorganisms were discovered in our environment and linked to typhoid fever and other diseases that have plagued humanity, public health authorities have been concerned with the accumulation of filth and foul odours in urban areas. The first early inspection systems based on sensory evaluations were legally enforced at the beginning of the 20th century. Initial bacteriological techniques to detect pathogenic bacteria in foods, such as shellfish, appeared soon after. From that point on, the food and beverage industry has applied stricter product inspection procedures and more and more effective production methods to conserve the freshness of natural raw materials. Today, the establishment of good manufacturing practices (GMP) and good hygienic practices (GHP) in many countries has significantly reduced the risk of spoilage and pathogenic microorganisms in modern food products. In addition to complying with national and international food regulations, food manufacturers are required to follow international quality standards, such as ISO as well as the Hazard Analysis Critical Control Point (HACCP) system.

In recent years, there has been an increasing focus on traceability in food production. This has followed public concerns arising from cases of food contaminations and the development of foods containing ingredients derived from genetically modified (GM) crops. In the light of the increasing need for food more rapid food testing, it became clear that, the traditional microbiological detection and identification methods for food borne pathogens was no longer effective, because, it was time consuming and laborious to perform, and are increasingly unable to meet the demands for rapid quality control. A rapid method is generally characterized as a test giving quicker results than the standard accepted method of isolation and biochemical and/or serological identification. Some of the newer and more rapid methods of food quality control are:

Ion Mobility Spectrometry (IMS) or Differential Mobility Spectrometry (DMS)

These methods are used in the identification and quantification of analytes with high sensitivity. The selectivity can even be increased - as necessary for the analyses of complex mixtures - using pre-separation techniques such as gas chromatography or Multi-Capillary Columns (MCC). The method is suitable for application in the field of food quality and safety -including storage, process and quality control as well as the characterization of food stuffs.

Electronic-nose

This is an instrument which comprises an array of electronic chemical sensors with partial specificity or broad-band chemical selectivity and an appropriate pattern recognition system, capable of recognizing simple or complex odours. This method is used to detect the bacterial growth on foods such as meat and fresh vegetables. It can also be used to test the freshness of fish. It is used in the process control of cheese, sausage, beer, and bread manufacture as well as for detection of off-flavors in milk and dairy products. This technique has other applications too. The advantages of the electronic nose can be attributed to its rapidity, objectivity, versatility, non-requirement for the sample to be pre-treated and ease of use Natale researcher.

Immunochemical Methods

This method is based on antigen – antibody interaction. The antibodies are highly specific for the antigen (analyte), and secondly, the antigen, the antibody, or an antiglobulin may be conjugated to an enzyme that produces an intensely colored or fluorescent product in the presence of the enzyme substrate to enhance the detectability of the analyte in an amplification step. Toxins produced by E. coli, Clostridium, Salmonella and Shigella have also been similarly detected.

Enzyme Immunoassay (EIA)

Microorganisms are often characterized and identified by the presence of unique protein carbohydrate markers also called antigens, located within the body or the flagella of the cell. Detection of these unique antigens has been a cornerstone of diagnostic microbiology for many years. In recent years, EIA using monoclonal antibodies have made available rapid and consistent microbiological detection systems. The most widely used systems employ a sandwich technique using antibody attached to a polystyrene matrix to which the sample is added. Post incubation, a second antibody, which is specific for the organism and has been tagged with an enzyme, is added. The addition of enzyme substrate to the mixture completes the EIA. The presence of the specific organism results in a colorimetric change in the enzyme substrate, which may be observed visually or with a spectrophotometer. Most EIA are very specific but lack sensitivity. Normal sensitivity has been reported to be in the range of 106 org/ml.

Biosensors

Biosensor is usually a device or instrument comprising a biological sensing element coupled to a transducer for signal processing. Biological sensing elements include enzymes, organelles, antibodies, whole cells, DNA, and tissues. There are different types, conductance bioluminescence enzyme sensors utilizing potentiometric, amperometric, electrochemical, optoelectric, calorimetric, or piezoelectric principles. Basically, all enzyme sensors work by immobilization of the enzyme system onto a transducer. This technique provides sensitive, miniaturized systems that can be used to detect unwanted microbial activity or the presence of a biologically active compound, such as glucose or a pesticide in food. Immunodiagnostics and enzyme biosensors are two of the leading technologies that have had the greatest impact on the food industry.

Flow Cytometry (FCM)

Specific detection of pathogenic strains can be achieved by Flow cytometry using immunofluorescence techniques, which allow microorganism detection at the single-cell level. Although this technology can be used for food samples, it requires prior isolation of the target organism to generate antibodies. FCM finds wide application in milk and brewing quality control. The advantage of FCM is that it can also differentiate Viable Non-Culturable (VBNC) form of bacteria from healthy cultivable cells. This technology has the ability to detect microorganisms at relatively low concentrations in a short time, while multiple labeling allows the detection of different organisms or different stages in the same sample.

Polymerase Chain Reaction (PCR)

This technique involves the following steps: Isolation of DNA from the food, amplification of the target sequences, separation of the amplification products by agarose gel electrophoresis, estimation of their fragment size by comparison with a DNA molecular mass marker after staining with ethidium bromide and finally, a verification of the PCR results by specific cleavage of the amplification products and by restriction endonuclease or southern blot. Alternatively amplification products may be verified by direct sequencing or a second PCR.

Pulsed-field Gel Electrophoresis (PFGE)

PFGE is a restriction-based typing method that is considered by many to be the "gold standard" molecular typing method for bacteria. In this electrophoretic approach, DNA fragments are separated under conditions where there is incremental switch of the polarity of the electric field in the running apparatus. This technique allows for the resolution of DNA fragments up to 800 kb in size. When DNA is restricted with a restriction enzyme, PFGE provides a DNA "fingerprint" that reflects the DNA sequence of the entire bacterial genome. PFGE is a widely accepted method for comparing the genetic identity of bacteria. PFGE typing has demonstrated a high level of reproducibility for food borne pathogens. A major advantage of this method is its universal nature making it useful in bacteria sub typing, however, its limitation is that it is time consuming.

Magnetic Separation

One of the typical applications of this technique was reported by Mattingly & Safarik. The researcher separated salmonella from food and faecal matter using myeloma protein and hybrid antibody (for O antigen), conjugated to a polycarbonate- coated metal bead. It has also been reported that, Food sample like milk, yogurt, meat and vegetables can be tested. The challenge is in detecting E. coli is in the isolation of pathogenic strain from nonpathogenic strains. Immunomagnetic detection of listeria monosytogens has also been reported by Skjerve researcher. While Johne et al using magnetic beads coated with polyclonal antibodies was able to detect and isolate the specific protein of s. aureus.

Near-infrared (NIR) Spectroscopy

NIR has proven to be an effective analytical tool in the area of food quality control. The key advantages of NIR spectroscopy are (1) its relatively high speed of analysis, (2) the lack of a need to carry out complex sample preparation or processing, (3) low cost, and (4) suitability for on-line process monitoring and quality Control. The disadvantage of this method includes the requirement for large sample sets for subsequent multivariate analysis. Recently, researchers at Zhejiang University in Hangzhou, China used Vis/NIR spectroscopy together with multivariate analyses to classify nontransgenic and transgenic tomato leaves.

X-ray

This is a relatively newer technology in food quality control. X-rays started making in-roads into the food industry in the early 1990s. The driving force behind this was the increasing number of foreign bodies which could not be identified by metal detectors. Other than contaminants like glass, bone, rubber, stone or plastic, some specific applications are also more challenging for metal detectors, such as fresh meat and poultry, or foil-wrapped products. X-ray inspection has considerable advantages in many foods and beverage processing environments in that, it is easy to install, safe and simple to use, even without previous experience. It quickly and consistently identifies substandard products, reducing product recall, customer returns and complaints, therefore protecting manufacturers' brands and most importantly, preventing ill health.

Computer Vision

Aguilera et al reported a computer system that consisted of four basic components: the illumination source, an image acquisition device, the processing hardware, and suitable software modules. Their study was focused on analyzing the relevance of computer vision techniques for the food industry, mainly in Latin America. The researcher described how the use of these techniques in the food industry eliminates the subjectivity of human visual inspection, adding accuracy and consistency to the investigation. They also reported that, the technique can provide fast identification and measurement of selected objects, classification into categories, and color analysis of food surfaces with high flexibility. They opined that, since the method was non-contact and non-destructive, temporal changes in properties such as color and image texture can also be monitored and quantified.

Quality and Specification of Raw Materials

The concept of food safety covers all the steps involved in food chain starting from cultivation to consumption or "farm to fork." Good hygienic practices are therefore important to be followed from the procurement of the raw resources, transport of raw food, storage of raw food materials more so for commercial establishment.

In earlier times when the food safety concept was not so critically emphasized. There was dichotomy in research and development activities with each team working in isolation. But now convergence of all teams put in their expertise due to increase access to unique and complex materials, global sourcing handling methods, customer locations, and regulations. The unified team assess, if the material has limitations or may be too costly to handle, and determines if additional measures are necessary to prevent potential safety issues for the employees and product.

Components of Food Processing

Raw materials, including ingredients, processing aids, and packaging, are the foundation of finished food products. As such, they must meet not only acceptable and standard specifications, but also regulatory requirements such as safety, legal requirements. The specifications must be such that they contribute to the functionality and quality of the chosen process and the product.

Raw Material Selection

Appropriate raw materials must be selected based on functionality. Functionality encompass multiple requirements based on identified characteristics of the finished product (binders, thickeners, type of resin for plastic packaging, etc.), organoleptic characteristics (flavor, color, aroma, texture), product safety characteristics (to lower the pH or water activity), and preservatives (extension of shelf life, color, or flavor retention, etc.).

No raw material or ingredient should be accepted by an establishment if it is known to contain parasites, undesirable microorganisms, pesticides, veterinary drugs or toxic, decomposed or extraneous substances which would not be reduced to an acceptable level by normal sorting and/or processing. Where appropriate, specifications for raw materials should be identified and applied.

Raw materials or ingredients should, where appropriate, be inspected and sorted before processing. Where necessary, laboratory tests should be made to establish fitness for use. Only sound, suitable raw materials or ingredients should be used.

Raw Material Specifications

The key characteristics for the raw ingredients must be documented with all the relevant specifications. This may be the dated Technical Data Sheet from the supplier. A purchase order should list the supplier's specific name and item number for the material. It is good to keep scope to revise this data sheet as per periodic need. The manufacturer should obtain a certificate of analysis for each lot (code). Each incoming lot should be sampled according to a predetermined sampling plan and analysed for adherence to specifications.

Flexibility in Sourcing and Cost

Is the material a commodity-type item that can be purchased from multiple sources? If so, compare existing specifications from multiple suppliers; allow comparison bidding/purchasing. Can the tolerances for characteristics be expanded to be able to purchase from more than one source or a wider range of possible, existing materials (such as granulation size for materials that are going to be dissolved or melted)? Can purchasing find a similar functionality material that costs less or has fewer concerns?

Considerations in Selection

If there is a raw material already in use with similar characteristics it can be retained. Depending on whether it is agricultural item, commodity item or one with standard identity general specification that can be used between multiple suppliers can be developed and maintained. The raw materials must fall under Generally Recognized as Safe (GRAS) listing as approved by the food regulatory body of the country. Ban on the use of as genetically modified materials depends on the legal status on use of these items. Maximum levels for use, both in the country of manufacture and the country of sale/use again depends on the food laws of both the countries. The raw material must also comply with the existing company or customer standards namely for e.g., kosher, halal, organic, gluten-free and so on. Specification requirements can be documented and recorded.

Raw Material Storage – Refrigerated and Non-refrigerated

a) All raw materials should be:

- Free from insect and rodent infestation.

- Free of contamination from other sources, e.g. birds, moisture, mold etc.

If bulk floor handling and storage is in use then,

- Hose couplings, inside and outside plant, should be adequately protected from rodents, clean and in good repair.
- Dust collectors or ventilation bags at top of the bulk tank should be clean and insect free.
- If system contains inspection ports, they should be free from contamination.

b) The microbial load of all the food items like raw fruits and vegetables, liquid milk, meat, eggs, flour, cereal grains etc. should be within the acceptable limits at receiving as well as during their storage in the raw material store.

c) Depending on the nature of the food adequate facilities shall be available for storing refrigerated or frozen foods, monitoring food temperatures, and controlling ambient temperatures and humidity to ensure the microbiological safety and suitability of food.

d) In case of grain storage, fumigation should be done at regular intervals to ensure the absence of any microbiological and pest activity.

High Risk Raw Foods Sourcing and Storage

Milk and milk based products, meat, fish, vegetables and fruits perish fast. These items must be free from contamination at source and must be later stored at proper temperature, humidity and other conditions necessary to keep them safe. Precaution must be taken to follow the hygienic principles while processing and preparing foods using these commodities.

Inventory Management

Stocks of raw materials and ingredients should be subject to effective stock rotation. Prevention of health hazards begins with control of incoming materials. Inadequate incoming ingredient controls could result in product contamination and/or under processing. The degree of control exercised over incoming ingredients is appropriate to the risk. The principle of first in first out should be followed.

The manufacturer should control incoming ingredients through the following programs. The first three options apply to ingredients that may be critical factors where further processing is not likely to eliminate a hazard. The fourth option applies to ingredients that are not likely to have an impact on the safety of the food.

A.Periodic evaluation of incoming ingredients B.the manufacturer should have written specifications for ingredients C. purchasing specifications should include a provision for compliance with the food law. D. The manufacturer should maintain a documented history of adherence to specifications for each supplier, e.g. analytical results.

Certificate of Analysis

The manufacturer should obtain a certificate of analysis for each lot (code). A statistically representative sample should be taken to verify the accuracy of the certificates of analysis at a scheduled frequency. A new history of adherence to specifications should be established when a firm changes suppliers, purchases ingredients from a new supplier or purchases a new ingredient from an existing supplier or when spot checks do not agree with the certificate of analysis.

Vendor Certification

When the manufacturer relies on vendor certification the following minimum requirements should be in place:

- The manufacturer should have written specifications for ingredients.
- The manufacturer should have documentation to demonstrate adequate knowledge of the vendor's process, e.g. process flow charts, on-site evaluations, identification of critical control points, specifications, control limits, monitoring programs and frequencies, corrective action and verification procedures.
- The manufacturer should have data to demonstrate the capability of the vendor's process to manufacture consistently within specifications.
- Prior to implementation of a periodic monitoring program the firm should analyse an appropriate number of consecutive lots to establish a historical database and confirm adherence to specifications.
- The manufacturer should conduct periodic monitoring to verify adherence to specifications.
- The manufacturer should conduct vendor audits to validate the status of the vendor certification program.

Occupational Safety and Health Administration (OSHA)

OSHA Considerations

The safety of materials such as toxic, inflammable or irritants need handling safely and the facility for the same must be available. The food handlers must know of the safe handling procedures.

Product Safety Considerations

Historical information about the raw materials can be obtained from the literature. The company can conduct HACCP to assess the potential biological, chemical, and physical hazards affiliated with the raw material (HACCP/HARPC review) as well as affiliated prerequisite programs and downstream prevention/elimination/reduction steps for identified hazards (physical, chemical, and biological). If there is a need for a new procedure (e.g., an allergen program and/or validation) over the existing procedure, it can be included or revised.

Facility and Equipment Capabilities

The plant must be technologically appropriate to handle the material. The machinery used should not change the characteristics of the ingredient, should be capable of handling the raw material of specific particle size, design wise suitable and not introduce process contaminants.

Water

As an Ingredient

Only potable water should be used for food preparation. Water should meet the requirements of the regulations. Water should be analysed by the manufacturer or municipality at a frequency adequate to confirm its potability. Water from sources other than municipal supplies must be treated as necessary and tested for potability.

Ice and Steam

Ice used as an ingredient or in direct contact with food should be made from potable water and protected from contamination. Similarly steam coming in contact with food should be made from potable water. For other purposes like water for use in fire control, washing and other purposes not connected with food directly, other sources of clean water may be used.

Water recirculated for reuse should be treated and maintained in such a condition that no risk to the safety and suitability of food results from its use. The treatment process should be effectively monitored. Recirculated water which has received no further treatment and water recovered from processing of food by evaporation or drying may be used, provided its use does not constitute a risk to the safety and suitability of food.

There should be no cross-connections between potable and non-potable water supplies. All hoses, taps and other similar possible sources of contamination should be designed to prevent back-flow or backsiphonage.

Storage of Water

If it is necessary to store water, storage facilities should be designed, constructed and maintained to prevent contamination. The volume, temperature and pressure of the potable water should be adequate for all operational and clean-up demands. Water if treated with chemicals, should not result in unacceptable chemical residues in the water. Chemical treatment should be monitored and controlled to deliver the desired concentration and to prevent contamination. Recirculated water should have a separate distribution system which is clearly identified.

Packaging

For packaging materials, the specific composition of the packaging material would be specified, such as glass, polyethylene (PET), polypropylene (PPE), and so forth. Packaging design and materials should provide adequate protection for products to minimize contamination, prevent damage, and accommodate proper labelling. Packaging materials or gases where used must be non-toxic and not pose a threat to the safety and suitability of food under the specified conditions of storage and use. Where appropriate, reusable packaging should be suitably durable, easy to clean and, where necessary; disinfect. Where appropriate, the manufacturer should control incoming packaging materials from the supplier using an approach similar to the options suggested for incoming ingredients.

Size and Type of Packaging

The size and type of packaging depends upon the nature of the food to be packed, quantity, transportation, time on shelf, single or multiple use and consumer preferences. The package should have adequate space for the required labeling as per the regulations. Following are the information required on packaged food:

The ingredient/material composition is listed in decreasing order of presence or as outlined in labelling regulations:

- The name of the product and general name.
- The supplier's item number.
- Components or composition of the materials.
- Information on the presence of regulated or customer-recognized food allergens.
- Organoleptic information (appearance, flavor, and aroma).
- Relevant physical, chemical, and microbiological information.
- Shipping and storage information.
- Shelf life.
- Handling information.
- Allergens if they are likely to be present.
- Instructions for use.
- Best before date.

Management and Supervision

The type of control and supervision needed will depend on the size of the business, the nature of its activities and the types of food involved. Managers and supervisors should have enough knowledge of food hygiene principles and practices to be able to judge potential risks, take appropriate preventive and corrective action, and ensure that effective monitoring and supervision takes place.

Documentation and Records

Where necessary, appropriate records of processing, production and distribution should be kept and retained for a period that exceeds the shelf-life of the product. Documentation can enhance the credibility and effectiveness of the food safety control system.

Critical records, e.g. records related to the adequacy of the thermal process and the achievement of a hermetic seal, should be signed and dated by a qualified individual designated by management prior to distribution of the product. All other records should be reviewed at an appropriate frequency to provide an early indication of potentially serious deficiencies. Records should be retained for one year after the expiry date on the label or container or, if there is no expiry date, for two years after the date of sale. Records should be maintained at the manufacturing plant and should be available.

Recall Procedures

Managers should ensure effective procedures are in place to deal with any food safety hazard and to enable the complete, rapid recall of any implicated lot of the finished food from the market. Where a product has been withdrawn because of an immediate health hazard, other products which are produced under similar conditions, and which may present a similar hazard to public; health, should be evaluated for safety and may need to be withdrawn. The need for public warnings should be considered.

- Each pre-packaged food should have permanent, legible code marks or lot numbers and where required expiry or "best before" dates on the package.

Recall Capability

The manufacturer should be capable of producing accurate information on a timely basis to verify that all affected product can be rapidly identified and removed from the marketplace. This can be demonstrated by the manufacturer as follows:

- Records of customer names, addresses and telephone numbers available for the lot tested.
- Records of production, inventory and distribution by lot available for the lot tested.

Product Inspection

Verifying the Product Quality and Consistency

Before a batch of product leaves the factory it must pass a 'positive release' test to confirm it is safe to consume. Quality tests have to be performed to verify product compliance with internal and external standards, including for harmful compounds or microorganisms in the materials used, the environment of operation and in products. Tests for allergens may have to be conducted if applicable.

Analytical Information

Food safety parameters or tolerances could include biological, chemical, or physical characteristics.

- Biological – Microbiological limits for pathogens, such as Salmonella and Listeria monocytogenes.
- Chemical – Fortification levels, sulfite levels, heavy metal content, etc.
- Physical – Size and foreign material (rocks, glass, metal, bones, etc.).

Functionality or quality parameters would include characteristics that can impact the functionality of the material or adversely impact your product.

- Biological – Microbiological limits for spoilage organisms or indicators of poor sanitation, including total plate count, yeast, mold, and coliform.
- Chemical – Characteristics such as concentration levels or purity.
- Physical – Characteristics such as viscosity, color, granulation size, insect parts, crush strength, physical measurements, etc.

Transportation

For chilled or frozen products, this also means ensuring they are stored and transported at the correct temperature. The transportation mode and capacity to be used depends again on product specifications. The following measures must be followed to ensure food safety during transportation:

- Food must be protected from potential source of contamination.
- Protect food from damage that may render it unfit for consumption.
- Provide an environment which effectively controls the growth of pathogenic or spoilage microorganisms and the production of toxins in food.

If these aspects are not taken care during food transit it may get contaminated, or may not reach its destination in a suitable condition for consumption, even if hygiene control measures have been taken earlier in the food chain.

Container Requirements

Conveyances and bulk containers should be designed and constructed so that they:

- Do not contaminate foods or packaging.
- Can be effectively cleaned and disinfected.
- Permit effective separation of different foods or foods from non-food items during transport.
- Provide effective protection from contamination, including dust and fumes.
- Can effectively maintain any temperature, humidity, atmosphere and other conditions necessary to protect food from harmful or undesirable microbial growth and deterioration likely to render it unsuitable for consumption.
- Allow any necessary temperature, humidity and other conditions to be checked.
- Ingredients requiring refrigeration should be transported at 4 °C or less and should be appropriately monitored. Frozen ingredients should be transported at temperatures that do not permit thawing.

Use and Maintenance of Containers

Conveyances and containers for transporting food should be kept in an appropriate state of cleanliness, repair and condition. Where the same conveyance or container is used for transporting different foods, or non-foods, effective cleaning and disinfection should take place between loads.

Where appropriate, particularly in bulk transport, containers and conveyances should be designated and marked for food use only and be used only for that purpose. Carriers should be inspected by the manufacturer on receipt and prior to loading to ensure they are free from contamination and suitable for the transportation of food. The manufacturer should have a program in place to demonstrate the adequacy of cleaning and sanitizing. For example, for bulk carriers a written cleaning and sanitizing procedure should be available. Carriers should be loaded, arranged and unloaded in a manner that prevents damage and contamination of the food.

HACCP plans developed by the food industry must consider the control of temperatures and contamination during the transport of foods. A food business operator may require an HACCP plan as a condition of doing business with a particular transportation firm.

Ultrasonic Technology to Improve Food Quality

Improving the nutritional values and stability of quality is a very important parameter in food product quality assurance for a healthy life of human beings. Consumers are looking for fresh and good characteristics in their food with nutrient content and high sensorial quality. Now, consumers are more aware of the processing techniques used in the processing of their food, and they prefer natural products free of additives and chemicals. Therefore, there is a need for alternative technologies for food processing. Recently, various modern thermal and non-thermal technologies such as pulsed light, pulsed electric field, high and low hydrostatic pressure, microwave, ohmic heating, freezing, pasteurizing, ionizing radiation, etc. have been used to improve the physicochemical characteristics, extend the shelf life of food products, and control food quality by inactivating microorganisms at sublethal or ambient temperatures. One of the non-thermal technologies that can be used also is the application of ultrasonic (high-power and low-power ultrasonic with low and high frequency); especially it has shown a negligible effect on the nutrient value of food products. Applications of ultrasonic technology for food processing aim to offer consumers high-quality foods. The ultrasonic is considered to be a promising and emerging technology that can be used in food processing technology and many industrial applications by regulating frequency. According to sound wave ranges used, the ultrasonic can be divided into low-power high-frequency ultrasonic and high-power low-frequency ultrasonic. Low-power ultrasonic with high frequency is used for non-destructive quality evaluation of physicochemical characteristics of fruit, vegetables, and food products during processing or storage.

The high-power ultrasonic with low frequency is used to improve the physicochemical properties of food products and in food processing such as humidification, hydrothermal treatments, extraction, drying, freezing, and inactivation of microorganisms of food products. The ultrasonic technology has been also used in the industry of food products to develop many reliable and effective processing applications of food. The most common applications of ultrasonic in the industry of food include extraction of intracellular and material cell destruction. Depending on the ultrasonic intensity, the ultrasonic is used for the deactivation or activation of enzymes, homogenization and mixing, dispersion, stabilization, crystallization and dissolution, emulsification, hydrogenation, preservation, ripening, meat tenderization, oxidation, as a solid-liquid extraction adjuvant to accelerate and to improve the extraction, and atomization and degassing of food processing. The objectives of ultrasonic research are to analyze and study the phenomena of undesirable and desirable degradation resulting from the applications of ultrasonic wave treatments in foods. The processing using ultrasonic may impact the chemical composition texture of foods.

Generally, ultrasonic applications are environmentally friendly and offer an advantage in the selectivity, yield, and productivity, with enhanced quality, reduced physical and chemical hazards, and short processing time. Before the commercialization of some food products such as vegetables and fruit, oils and fat, cocoa-sugar and coffee, meal and flours, dairy, and meat which are complex mixtures of proteins, sugars, lipids, vitamins, aromas, fibers, antioxidants, pigments, and mineral and organic compounds have to be processed and preserved using ultrasonic applications for food meals and to extraction of food ingredients.

Overview of Sound Waves

The sound wave type is determined by its frequency. Figure shows the sound spectrum which displays the various frequencies present in a sound. "Infrasound" indicates a sound wave below the human hearing range. This frequency of sound is used by submarine sonar devices and whales. The frequency of the sound for the human hearing ranges from 20Hz to 20 kHz. The sound signal arises from many sources, e.g., the air turbulence or gases, passage through fluids, and by the impact of solid against another solid similar or non-similar. Because the sound is a natural phenomenon of waves, it may contain only one frequency as a sine wave with pure steady state or contain complex frequencies such as the noise generated by many sound sources, e.g., machines and engines. The frequency of sound (f) is sound pressure times number. The sound frequency also may be identified by the frequency of angular (ω) expressed in radians per second as shown in Equation $\omega = 2\pi f = \frac{2\pi}{T}$. The period (T) is the time amount for a cycle of the single:

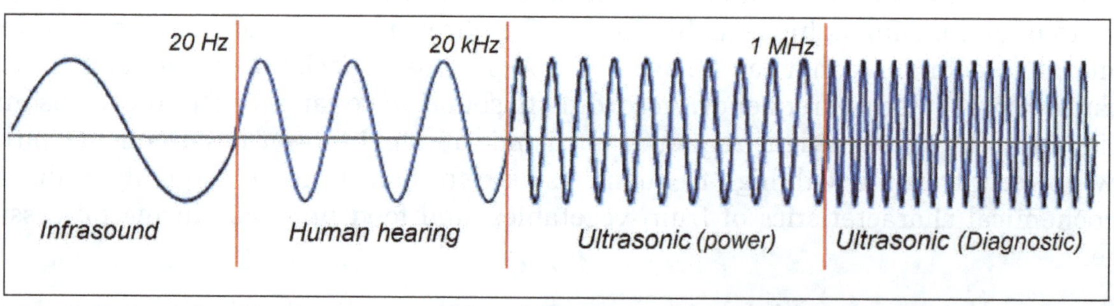

Sound spectrum.

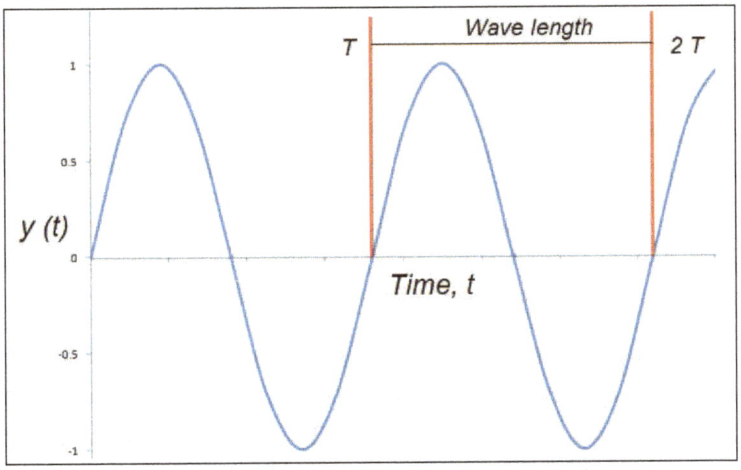

The pure steady-state sine wave pulses.

Actually, the amplitude of the sound wave is strongly affected by the particles near the source of the sound waves, and on the contrary, the deeper particles are in the treated medium, the lower the sound wave amplitude. This reduction in sound wave amplitude at the deep is due to the attenuation produced by the treated medium. As a result, the sound amplitude versus wavelength distance is actually an exponentially sinusoid degenerate. The wavelength ((λ) is the distance between peaks of successive amplitude) is related to frequency (f) through the traveling wave velocity (c) as shown in Equation below:

$$\lambda = \frac{c}{f}$$

Ultrasonic is a wave of sound with a frequency greater than the human hearing limit. Ultrasonic is considered an energy form generated by a longitudinal mechanical wave with one-dimensional propagation and frequency of vibration above 20,000 cycles per second (20 kHz) as shown in Figure. Ultrasonic waves can be categorized according to its frequency into two categories that are: (1) Low-frequency category which has frequency ranging from 20 to 1000 kHz. The applications of this category are used at high-power intensities in industrial applications, ultrasonic therapy, sonochemistry, and nanotechnology. (2) High-frequency category which has a frequency above 1 MHz and is being used at low-power intensities for non-destructive quality evaluation, imaging, and diagnostic applications.

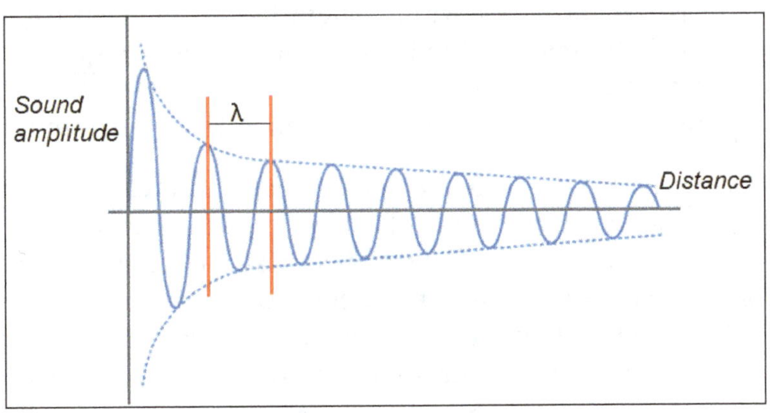

Sinusoidal ultrasound wave.

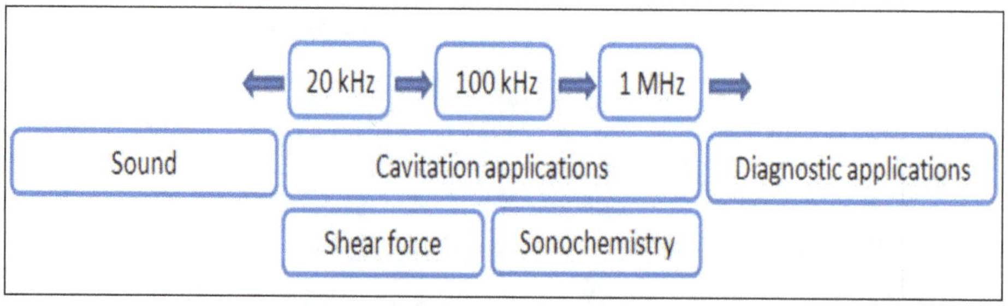

Ultrasonic frequencies classification.

Use of ultrasonic application provides a good way to reach higher rates for the chemical and physicochemical process, shorter processing times and pathways of reaction. Interaction mechanisms between the product material and ultrasonic waves vary as a function of the input power of the ultrasonic. The pulse of ultrasonic speed depends on the acoustic properties of the medium of treated material. The speed of sound propagation in solid materials is higher than the sound propagation speed in liquids and greater in liquids than in gases.

Ultrasonic Equipment

The main equipment of ultrasonic consists of a transducer, electrical power generator, and sound emitter devices. The emitter's function is to physically send the waves of ultrasonic to the medium. There are two types of ultrasonic systems used in the industry of food products: one using the bath as a traditional method and other using the horn as the sound emitter. The horn-based system is utilized in many applications from ultrasonic application in food processing and cleaning of plant surfaces for the process of food to application of ultrasonic for welding of metals.

Ultrasonic Transducers

The transducer is the most important part of ultrasonic systems; the role of the transducer in the system is to generate the actual ultrasonic waves by converting the mechanical or electrical energy into sound energy at ultrasonic frequencies by vibrating mechanically. The ultrasonic transducer contacts to an electrical generator with 20 kHz frequency to transform electrical energy into ultrasonic energy by mechanical vibration at the same frequency (20 k cycles per second). The most applicable methods of ultrasound generation are carried out using ultrasonic transducers depending on the principle of the electrostrictive transformer. The principle of the methods is based on ferroelectric materials' elastic deformation within a high-frequency electrical field which results in molecules' polarized mutual attraction in the field. Then, the high-frequency alternating current is transmitted via two electrodes to ferroelectric material. After generating mechanical oscillation, the waves of sound are transmitted to the amplifier to generate the ultrasound.

The ultrasonic transducer is an electronic device that generates and receives the waves of sound. The transducer basically functions as a converter of energy, where it converts a form of acoustical energy into other energy forms (e.g., mechanical, electrical, or thermal energy). In addition, the transducer is reversible in either direction to convert electrical or mechanical energy to sound energy or vice versa. The most high-intensity ultrasonic generators are essentially magnetostrictive devices crystal oscillators in use. The categories of ultrasonic transducers fall into the following:

- Crystal oscillators are work through the effect of piezoelectric (reversible).

- Magnetostrictive equipment are works based on the phenomenon of magnetostriction (reversible also).
- Mechanical transducers that operate as generators and receivers.
- Electromagnetic transducers are work based on the principle of the audio loudspeaker (but only work in the lower frequencies range of ultrasonic).
- Other different types are thermal, optical, and chemical transducers.
- Ultrahigh transducers that operate in the frequency range at megahertz or gigahertz.

Generally, the main transducers used in the most ultrasonic application can be summarized into three types: piezoelectric, magnetostrictive, and liquid-driven. The piezoelectric and magnetostrictive transducers convert magnetic and electrical energy into ultrasonic energy. The liquid-driven transducers depend on mechanical energy to generate ultrasonic energy.

Electrical Generator

The electrical generators are used to supply the ultrasonic systems with the required electrical energy to drive the transducer. Generally, the electrical generator produces a suitable power rating for the ultrasound system and allows the power to be set only indirectly through current (I) and voltage (V) settings. The current represents the electron charge traversing an area over some time interval and measured in amp, the voltage represents the stored energy in the electrons and measured in volt, and the electrical power is the output of current and voltage. Electrical generators that are designed and operate in the low frequency ranged from 10 to 40 kHz for ultrasonic generally focusing on industrial therapeutic applications, welding, cleaning, and disinfecting applications.

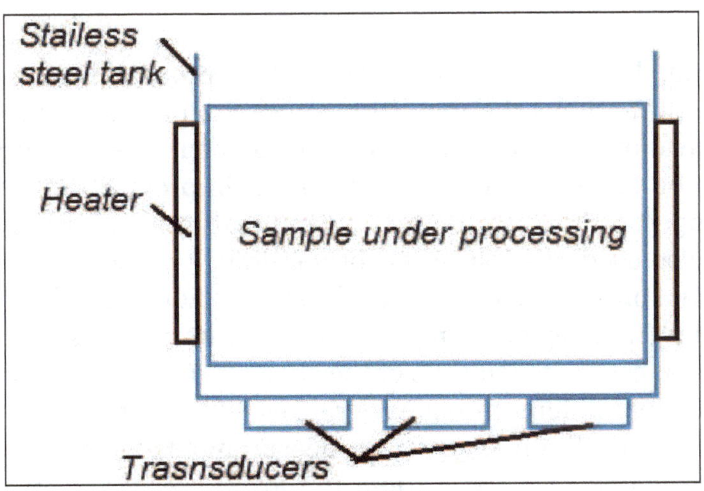

Ultrasonic bath.

Emitter (Reactor)

The function of the emitter (reactor) is to radiate the waves of ultrasonic which are produced by the transducer into the treated medium. In addition, the role of the emitter may also be to amplify the ultrasonic vibrations when radiating them in some ultrasonic system. The main types of emitters are horns and baths; the horns often require a sonotrode to attach with the horn tip. The baths usually consist of a stainless steel tank fixed with its base one or more transducers. The stainless steel tank holds a liquid case sample, and the transducers radiate ultrasonic directly into the sample.

Applications of Power Ultrasonic in Food Processing

Although the ultrasonic has been used in the twentieth century, most of the new and improved ultrasonic applications has reached practically only in the past few years. Ultrasonic applications can be classified into two categories as high intensity and low intensity. High-intensity applications deliberately affect the contents of the propagation medium. Uses of high intensities include liquid atomization, material machining, medical surgery and therapy, material cleaning, plastics and metals welding, biological cell disruption, and material homogenization. Low-intensity applications carry the objective of transmitting energy through a medium in order to convey information through the treated medium or to obtain information about the medium. Uses of low intensities include non-destructive testing, medical diagnosis, elastic property measurements of materials and agricultural products, and acoustical holography. Nowadays, ultrasonic application technology has extremely affected the meat industry, with a controlling role in the classification of the product quality. It is being used to measure the fat layer thickness in live animals, and it is also utilized to predict carcass traits as a livestock management part, and it has been used to improve homogenized milk quality. In addition, the ultrasonic application technology is utilized in the pest control that includes the expulsion or killing of insects. The potential uses of ultrasonic applications technology for improving the nutritional and quality aspects of food have been highlighted by Ashokkumar. The ultrasonic application technology offers a huge potential to bioprocessing industries and foods. Developing custom-made and new equipment is an issue to be addressed by food technologists, physicists, and engineers.

In addition, ultrasonic applications have been used for food processing as an important alternative processing method of conventional thermal. Ultrasonication process can preserve and pasteurize food products by inactivation of microorganisms and many enzymes at normal conditions of temperature to guarantee the safety and stability of foods for improving food quality. The changes in ultrasonic physical properties, such as attenuation and scattering caused by treated food product materials, have been also used in applications of food quality assurance. The potential applications of ultrasonic are not only affected by the medium (gas, liquid, solid, or supercritical) but also the treatments variables (flow regime, temperature, ultrasonic intensity, etc.) and the structure of product which could affect the magnitude of the changes induced by ultrasonic processing. Ultrasound can be divided into different frequency ranges. Most ultrasonic applications in the food processing technology involved non-destructive measurements which referred especially to the assessment of product quality; such applications use low power less than 1 W/cm^2 and high-frequency ultrasonic of 100 kHz to 1 MHz. Low-intensity ultrasonic is commonly applied as an analytical method to provide information on the food product's physicochemical properties such as acidity, ripeness, firmness, content of sugar, etc. The high power levels used (typically in the range 10–1000 W/cm^2) with low frequency (16–100 kHz) are used to make physical or chemical changes in the food to improve its properties.

Generally, the ultrasonic applications are separated into two categories: the first category is low-intensity ultrasonic (called non-destructive or high-frequency ultrasound), and the second category is high-intensity ultrasonic (called low-frequency or power ultrasound).

Applications of Low-intensity Ultrasonic

Low-intensity high-frequency ultrasonic is a non-destructive technique which is applied for detection purposes and provides information about the physicochemical characteristics of food products such as structure, firmness, composition, flow rate, physical state, etc. The action of ultrasonic waves is dependent on the input power. So the low-power ultrasonic is considered a non-invasive non-destructive method, and it is a useful technique for characterizing the physicochemical properties of food products, determining the food components type and contents, and measuring the emulsions droplet size. Irradiation of food products by low-power ultrasonic did not create any physical changes, at variance the high-power ultrasonic created the changes. It is also used as a processing method in the industry of food to describe the components of food products, often in line with quality assurance. The non-destructive test basically is done by sending waves of ultrasonic through the medium without causing any permanent electrical, chemical, or physical changes in the food products. This is due to the use of too low ultrasonic intensity ($<1\ W/cm^2$), so there is no change in the foods by using this.

When ultrasonic waves pass through the medium, the particles in the medium oscillate mechanically in response to the low-intensity (low-energy) ultrasound. After that, the particles exposed to the waves of ultrasonic simply return to their position of equilibrium when the ultrasonic source is stopped. The distance to the location of reflection can be calculated by measuring the attenuation coefficient and frequency properties of ultrasonic to evaluate the physicochemical properties and to allow detection of compositional changes in the food products. In using low-intensity ultrasonic to characterize vegetable and fruit properties, there must be a relationship between the property to be measured and any measurable parameter of ultrasonic (e.g., impedance, attenuation, or velocity). The particular parameter that often influences the properties of ultrasonic in vegetables and fruits is the presence of intercellular air spaces that causes a resonant phenomenon over ultrasonic frequencies in a wide range. The appropriate frequency which transmits normally through vegetables and fruits is above 1 MHz at low intensity to avoid the damage in plant tissue.

On the other hand, there are other indirect applications for high-frequency ultrasonic in food processing area such as applications of ultrasonic in humidifiers or misting devices which are used in humidification or hydration of fresh fruit and vegetables or humidification systems of meat in the cold storage rooms for improving the quality of the product and decreasing the weight loss during the storage period. The operating principle of ultrasonic humidifiers depends on converting the electrical energy into periodically mechanical vibration by piezoelectric transducers and horn, which vibrates at high frequencies. The piezoelectric transducers are placed at the bottom of the water in order to produce high-frequency waves that propagate upward into the water. Then the ultrasonic wave rarefaction cycle causes cavitation; in addition, the water over the piezoelectric transducer will produce a wavy layer. If the ultrasonic waves have enough energy that can overcome the water surface tension, then droplets will be generated from the water top surface. When the vibrating surface amplitude is increased to a level that the ultrasonic waves collapse and are unstable, the droplets will be ejected away from the water surface into a mist. The droplet's size is dependent on the frequency of vibration and water depth above the piezoelectric transducers.

The diameter of the atomized droplets is calculated based on the properties of the ultrasonic generator by Equation below:

$$D_d = \alpha_g \sqrt[3]{8\pi \times \sigma_s / \rho_s \times f^2}$$

Where D_d represents the droplet diameter, f is ultrasonic frequency, ρ is a liquid density, σ is a surface tension, and α_g is a dimensionless constant ($\alpha_g \approx 0.4$).

Generally, low-intensity ultrasonic applications can invaluably improve quality control in food production and monitor the changes that occur during humidification, emulsifying, freezing, or drying of food products. Some food manufacturers use non-destructive ultrasonic applications to locate foreign particles such as organic residues, bacterial infections, or glass in solid and liquid food products during and even after food packaging. Low-intensity ultrasonic has been used successfully at ultrasonic wave frequency of 150 kHz as a non-invasive and non-destructive means of evaluating the commercial poultry egg quality at different conditions of storage using the velocity ultrasound phase within the material of eggs to recognize the differences between the aged and fresh eggs.

Low-intensity ultrasonic applications are considered one of the efficient tools for non-destructive quality evaluation of fresh fruits and vegetables. These applications are characterized as a reliable and fast technique for correlating fruit and vegetable properties and specific indices of quality with the different growth stages, after maturation, during storage, and after storage to be ready for marketing and consumption while ensuring its quality. Commercial application of ultrasonic applications will be useful to consumers and growers due to the public demand for high-quality and uniform agricultural products. High-frequency ultrasonic technique using a contact transducer of 100 kHz as a non-destructive tool to determine fruit quality of navel oranges was applied successfully after fruit harvesting with a high accuracy level. Water content and density of the fruit can be determined accurately regardless of the other physical properties such as maturity, size, and the peel uniformity by isolating the results section which relates straight to the fruit acoustic properties. There is a high level of correlation between orange firmness and the reflected energy quantity of ultrasonic. Using ultrasonic technique, substandard individual fruit can be identified and sorted to be discarded at any harvest time and during processing or in a storage room. On the contrary, the methods of traditional destructive can be applied only on a limited sample of fruit after harvesting. The measurements of the ultrasonic velocity (highfrequency) and attenuation which was conducted at 25 MHz on samples of mango juice showed a big variability with a maturity of fruit at picking and after picking at ripening stage in relation to texture of fruit, the content of total soluble solids (TSS), and changes in biochemical composition. Many research has been done on non-destructive applications of ultrasonic technologies in food processing, but further future research is needed in this area in order to develop new automated ultrasonic equipment.

Applications of High-intensity Ultrasonic

Applications of high-intensity ultrasonic or power ultrasonic are used to change the physical or chemical properties of food products as well as to promote the reactions of chemicals, produce emulsions, inhibit enzymes, disrupt cells, crystallization processes modification, etc. The use of high-intensity lowfrequency ultrasonic waves generating sonotrodes was initially proposed for cleaning, emulsification, and bacterial lysing. The high-intensity ultrasonic wave (high-power) equipment using sonotrodes operating was further developed for processes of chemicals (up to 6 kW). In recent, ultrasonic systems are developed to generate high mega-sonic frequencies of ultrasound (400 kHz) with a high power level (>100 W). Therefore, the high-intensity ultrasonic wave is suitable for many applications in food products.

The high-intensity ultrasonic fundamental effect on the fluid material is for effective hydrostatic pressure on the medium and the imposition of acoustic pressure. The acoustic pressure (P_a) is a sine wave dependent on the ultrasonic frequency (f), time (t), and the wave pressure amplitude at the maximum ($P_{a\text{-}max}$). The maximum wave pressure is proportional to the transducer power input:

$$P_a = P_{a\,max} \sin(2\pi ft)$$

The application of high-intensity ultrasonic (high power level = 75 W) was developed and tested to assist in convective heat transfer during food drying. The application of ultrasonic is based on the ultrasonic energy transmission through airborne contacts and solid contact series between the ultrasonic transducer and the tray of the food product as a vibration surface of ultrasonic transmitting. The slices of apple were dried using this method without compromising the quality of the product. The results indicated that using the ultrasonic application during apple drying led to the following: processing time was accelerated, consumption energy was reduced, production throughput was increased, and the quality of the product was not affected by ultrasonic processing. The results also indicated that the ultrasonic treatments led to improve the convective drying process efficiency when using high-power ultrasonic at low temperature. These results are very useful at the need to dehydrate heat-sensitive products effectively or to decrease food drying time in order to preserve the physicochemical and nutritional properties of food products. Pasteurization of many food products by an ultrasonic application at 50 °C has a preserving potential on the food quality in terms of color, flavor and physicochemical properties compared to the techniques of conventional pasteurization at high temperatures. The propagation of ultrasonic in a medium causes chemical and physical impacts, and these impacts have been used to improve the efficiency of the operations of various food processing technologies, and it has been also used as diagnostic technology in food quality control. The high-intensity ultrasonic application was applied to control ice crystal's size distribution in low temperature processes and related applications such as thawing, freezing, freezedrying, and freeze concentration. It has been led to improve the freezing process efficiency, accelerate the freezing rate, and ensure frozen food quality.

High-intensity ultrasonic is being applied as an efficient preservation tool in fields of food processing for fruits and vegetables, honey, cereal products, proteins, gels, enzymes, cereal technology, dairy technology, water treatment, microbial inactivation, etc. The researchers have studied the effects of high-intensity ultrasonic at different levels of power ultrasound of 0, 200, 400, and 600 W as non-thermal processing on microbial inactivation (aerobic mesophilic, molds, yeasts, and coliforms), microstructure (particle size distribution and optical microscopy), rheology, color, and kinetic stability of the inulin-enriched whey beverage. The result obtained by applying ultrasonic power of 600 W was comparable to applying a high temperature of 75 °C at short treatment time of 15 s concerning the total microbial inactivation.

In addition, the high-intensity ultrasonic was better than the high-temperature short-time ultrasonic in improving kinetic stability of beverage, decreasing consistency and viscosity, avoiding phase separation, disrupting fruit and milk cells, and decreasing particle size. Therefore, non-thermal processing by high-intensity ultrasonic seems to be a promising technology for the production of probiotic dairy beverages. However, further future studies concerning the ultrasonic application effect on nutritional properties of this product must be evaluated before marketing.

Sterilization and improved emulsification can be conducted at lower temperatures than conventional treatments at high temperatures using high-intensity ultrasonic to produce a stable food product by retaining the useful bioactive ingredients and preventing spoilage of treated food. Applications of high-intensity ultrasonic in the fractionation of fat, dairy beverages production, and disruption of casein offer the potential of decreased treatment times; properties of the possible product have more advantages than those produced through conventional thermal techniques. Therefore, using ultrasonic applications in this area will lead to economic savings to producers in terms of producing value-added products and processing times and temperature. The consumers were satisfied with ultrasonic application studies for processing of food products to improve the quality of final products in terms of flavor, color, texture, and other physicochemical characteristics.

High-intensity ultrasonic treatment is a good process to inactivate enzymes and microorganisms at combined pressure and heat treatments as a hurdle technology. This combination is a successful application in lower temperatures for the inactivation process which provides a good solution for food product producers to secure fresh-like foods. The impact of power ultrasound on the fruit and vegetable quality during drying and pre-treatment has been assessed. The indicators of fruit quality such as the losses of leaching, rehydration capacity, shrinkage of fruit, and the final product's organoleptic characteristics have been also evaluated. The result showed that enzyme inactivation and leaching losses during blanching using high-intensity high-power ultrasonic at low temperature are similar to the result found using conventional treatments, but there is a significant reduction in the ultrasonic treatment time. Ultrasonic application in the drying of strawberries and carrots produces a highly significant reduction in the time of processing while providing high-quality final products. The final products' quality was equivalent or superior to final products obtained in convective dryer prototype under similar conditions, was higher than marketed products, and was similar to the produced products by freeze-drying. The impact of lowfrequency high-power ultrasound (40 kHz, 130 W) on bean in terms of kinetics of hydration and cooking times was studied. Treatment of bean samples by ultrasonic waves for 30 min at 30 °C 30 min occurs a significant increase in the effective diffusivity up to 45 times and reduces the time which obtains the equilibrium moisture content by 58.8% and the reduction percentage in cooking time reached 43%.

Generally, high-power ultrasonic has become an efficient technique for some commercial applications, such as homogenization, emulsification, crystallization, extraction, dewatering, low-temperature pasteurization, deforming, degassing, viscosity alteration, reduction of particle size, and inactivation or activation of enzymes. In addition, due to the need for inactivation of enzymes and microorganisms without destroying food nutrients, the high-power ultrasound applications are the best processing methods as a non-thermal alternative method to thermal processing treatments for food product preservation. This is due to continuous development and improvement in the design and manufacturing of ultrasonic equipment, but high-power ultrasonic for food processing like most innovative technologies in this field is not an effective technique for large-scale commercial application. Therefore, there is a need to conduct research on high-power ultrasonic for it to become an efficient large-scale commercial technology for processing food products.

Applications of Cavitation

The cavitation phenomenon (liquid rupture) is easily observed in water boiling, turbines, hydrofoils, and in seawater in the proximity of a rotating propeller of the ship. It happens in those liquid regions that are subject to rapidly vacillating pressures with high amplitude. Cavitations also happen in a liquid exposed to highenergy ultrasonic, considering that the sound travels through a small volume of fluid or water. During the negative half of the pressure cycle, the liquid is exposed to tensile stress, and during the positive half of the pressure cycle, the liquid is exposed to compression stress. Therefore, the bubbles entrapped in the liquid will extend and retract alternatively. When the amplitude of pressure is sufficiently large and the bubble initial radius is minimal than the critical value, Ro is given using the following equation:

$$R_o = 1/\omega\sqrt{3\gamma\left(\left(P_o + \frac{2T_{st}}{R_o}\right)^{/\rho}\right)}$$

where ω is a signal angular frequency, P_o is the hydrostatic pressure in a liquid, γ is a principal specific heat ratio of the gas in a bubble, T_{ts} is a surface tension at the bubble surface, and ρ is the intensity of liquid.

The sound pressure load (<10 Pa) exerted on the ear of human is very small, but the pressure of ultrasonic (MPa) in liquids can be high enough to create the cavitation phenomenon (can destroy the treated medium). Ultimately, the cavitations lead to free radical production and sonochemical that react chemically with media (liquid) and also lead to the destruction of microbiological cells. The ultrasound passage in liquid products generates a physical effect and mechanical agitation due to acoustic cavitation. The food industry has usually depended on the heating methods for enzyme and microorganism inactivation for preservation of food products. Despite thermal method actually leading to destroy some spores, kill microorganisms, and inactivate some enzymes, food may lose their organoleptic and nutritional properties during the process.

On the contrary, the inactivation mechanism using the ultrasonic application depends on the generation of physical forces due to the phenomenon of cavitation. Transmitted, dispersed, and reflected pulses of acoustic can be used in food product quality assurance. Using ultrasonic application for enzymatic inactivation of some food products is very important for the preservation of quality which is a requisite for secure food material stabilization. The physical and chemical forces generated by ultrasonic cavitation raise severe damage to the microorganism's cell wall, leading to microorganism inactivation. In addition, ultrasonic cavitation effects in liquid foods lead to disrupting the functional and structural components up to microorganisms cell lysis. The applications of ultrasonic that are used for flaw detection in food processing for quality assurance of food products must be designed with ensuring that no cavitation possibly occurs.

On the contrary, there are other applications of ultrasonic, depending on inertial cavitation to produce desirable changes in food products. These changes are produced by cavitation, such as microorganism inactivation and release of nutritional compounds and oils through the erosion of the cellular structure of the treated product cell. The released energy during cavitation has a great ability to improve food products' safety by destroying the pathogenic and food spoilage microorganisms and foreign material detection in food products. Although the applications of cavitation are well applied in many different industries other than food processing, the application of cavitation in processing of dairy products and its ingredients is recently gaining much attention, and it has a large potential to become a promising method in the near future in dairy product processing area such as reduction of viscosity, homogenization, making of yogurt, cream, and cheese, waste management, microbial inactivation, food safety, etc.

Power ultrasound cleaning application at low frequency generally operates between 20 and 50 kHz. The cleaning effect of ultrasound depends on cavitation. Increasing the cavitation in cleaning liquid increases the ultrasonic cleaning effect. The most important parameters affecting the cavitation are ultrasonic frequency and temperature.

Chapter 3

Methods of Food Preservation

Food preservation seeks to prevent the growth of microorganisms and in turn slow the oxidation of fats which can cause the food to become rancid. Some of the methods used for the preservation of food are refrigeration, canning, curing and pickling, and smoking. The topics elaborated in this chapter will help in gaining a better perspective about these methods of food preservation.

Low Temperature Preservation

The preservation of food by refrigeration is based on a very general principle in physical chemistry: molecular mobility is depressed and consequently chemical reactions and biological processes are slowed down at low temperature. In contrast to heat treatment, low temperature practically does not destroy microorganisms or enzymes but merely depresses their activity. Therefore:

- Refrigeration retards spoilage but it cannot improve the initial quality of the product, hence the importance of assuring particularly high microbial quality.
- Unlike thermal sterilization, refrigeration is not a method of 'permanent preservation. Refrigerated and even frozen foods have a definite 'shelf life', the length of which depends on the storage temperature.
- The preserving action of cold exists only as long as low temperature is maintained, hence the importance of maintaining a reliable 'cold chain' all along the commercial life of the product.
- Refrigeration must often be combined with other preservation processes (the 'hurdle' principle).

Food preservation at low temperature comprises two distinct processes: chilling and freez-ing. Chilling is the application of temperatures in the range of 0 °C to 8 °C, i.e. above the freezing point of the food, while freezing uses temperatures well below the freezing point, conventionally below 18 °C. The difference between the two processes goes beyond the differ-ence in temperature. The stronger preserving action of freezing is due not only to the lower temperature but also and mainly to the depression of water activity as a result of conversion of part of the water to ice. The use of refrigeration in the food industry is not limited to pres-ervation. Refrigeration is applied for a number of other purposes such as hardening (butter, fats), freeze concentration, freeze drying, air conditioning including air dehumidification and cryo-milling.

Effect of Temperature on Food Spoilage

The relationship between the temperature and the rate of chemical reactions is described by the Arrhenius equation:

$$k = A_e^{-\frac{E}{RT}} \Rightarrow \text{In}\, k = \text{In}\, A - \frac{E}{RT}$$

where,

- k - Rate constant of the chemical reaction.
- E - Energy of activation, j/mol.
- T - Absolute temperature, k.
- R - Gas constant.
- A - Frequency factor, almost independent of temperature therefore constant for a given reaction.

The kinetics of chemical spoilage reactions such as non-enzymatic browning and loss of some vitamins during storage is found to fit the Arrhenius model quite closely within a fairly wide temperature range. The Arrhenius model is therefore widely used for the prediction of chemical spoilage during storage, following non-monotonous time temperature profiles. Attention should be paid to the possibility of discontinuity in the Arrhenius model due to phase transition phenomena.

Effect of Low Temperature on Enzymatic Spoilage

The relationship between enzymatic activity and temperature follows the well-known bell-shaped curve, with maximum activity corresponding to the optimal temperature characteristic to each enzyme. On first sight, this behavior may seem to be in contradiction with the Arrhenius model. In reality, however, the bell-shaped behavior is the result of two simultaneous and contrary processes, both depressed by low temperature. The first is the enzyme-catalyzed reaction itself. The second is the thermal inactivation of the enzyme.

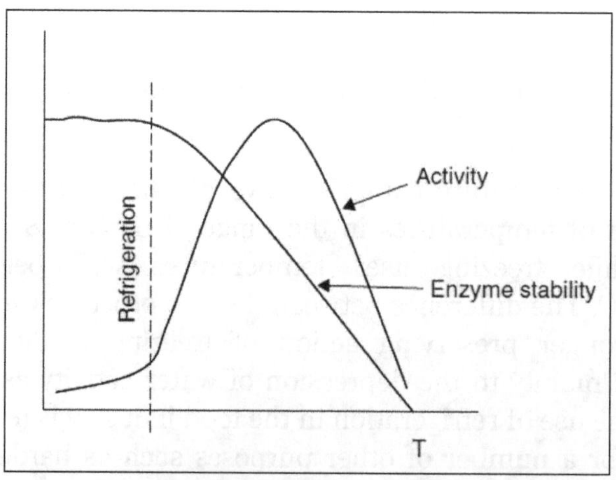

Effect of temperature on enzyme activity and stability.

Enzyme activity is strongly slowed down but not totally eliminated by refrigeration, even below freezing temperature. This is the reason for the need to inactivate the enzymes by blanching, particularly in frozen vegetables. Enzymatic activity in chilled and frozen foods is of considerable technological significance. Such activity may be desirable, as in the case of the aging of chilled meat or in the development of flavor in many kinds of cheeses but it can also be a source of deterioration such as the activity of proteases in fish or lipases in meat etc.

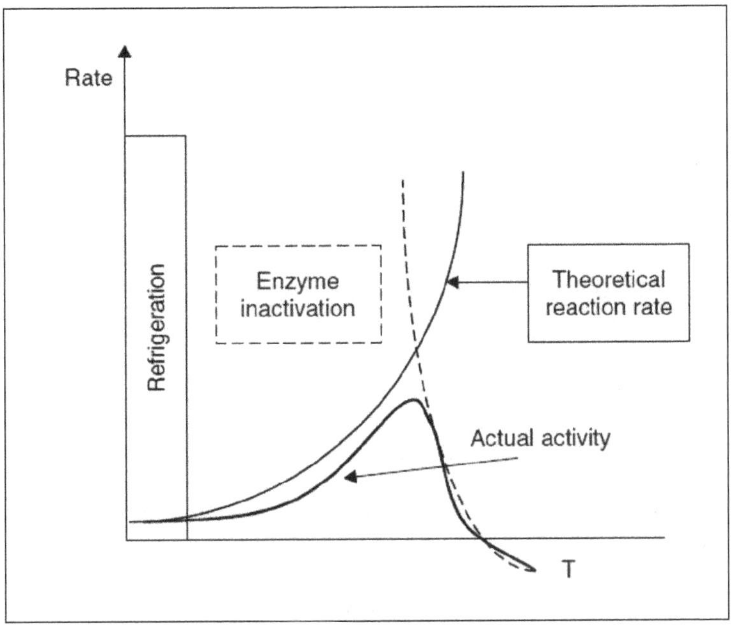

Bell shaped activity curve.

Effect of Low Temperature on Microorganisms

With respect to the effect of temperature on their activity, microorganisms are grouped into four categories: thermophiles, mesophiles, psychrotropes and psychrophiles. Typical growth temperature ranges corresponding to the four groups are given in Table. The four groups differ not only in their temperature requirement for growth but also in their rate of growth. Figure shows qualitatively the dependence of growth rate (generations per hour) and temperature for thermophiles, mesophiles and psychrophiles. Psychrophilic bacteria grow much more slowly, even at their optimal temperature. In refrigerated foods, psychrotropic and psycrophilic microorganism are obviously the main reason for concern.

Table: Bacteria, grouped by their growth temperature.

Group	Growth temperature (°C)		
	Minimum	Optimum	Maximum
Thermophiles	34 to 45	55 to 75	60 to 90
Mesophiles	5 to 10	30 to 45	35 to 47
Psychrotropes	– 5 to 5	20 to 30	30 to 35
Psychrophiles	– to 5	12 to 15	15 to 20

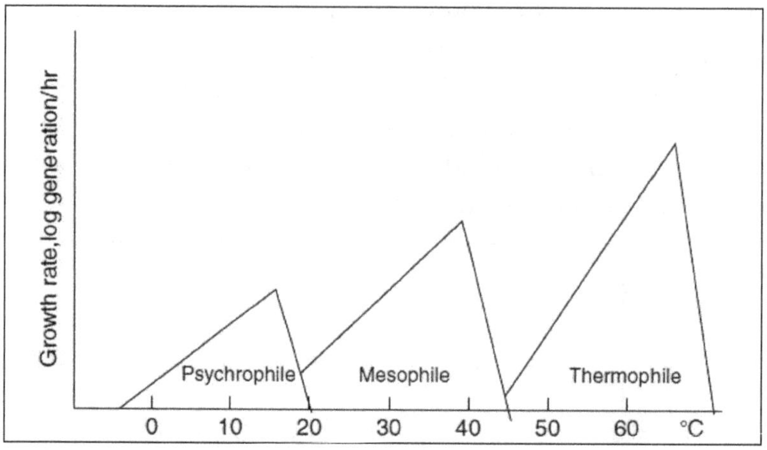

Classification of microorganisms by their response to temperature.

The relationship between temperature and shelf life is evident from the graph in Figure. It can be seen that, at lower temperature:

- The induction period (lag phase) is longer.
- Growth rate in the logarithmic phase is slower.
- As a result of the above, the bacterial count after a given storage time is considerably lower.

The microbiological quality of a refrigerated food depends therefore on the time temperature profile of the life of the product. Obviously, additional factors such as pH and water activity play an important role in the microbial processes in foods. Predictive microbiology is a new science, trying to develop tools for the prediction of the microbiological quality of foods, as a function of storage conditions.

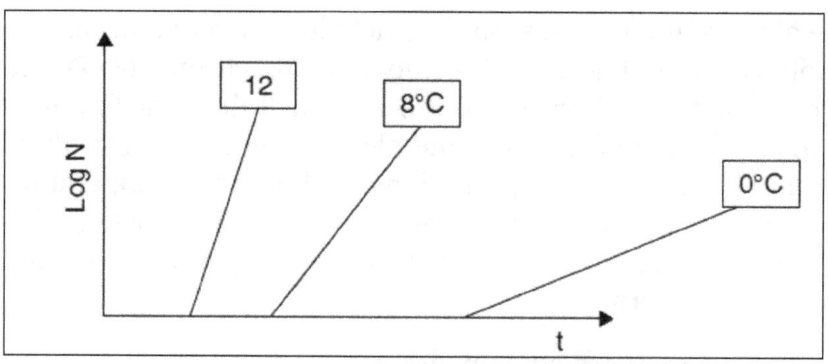

Schematic representation of the effect of storage temperature on the microbial load of a food.

Effect of Low Temperature on Biologically Active (Respiring) Tissue

By 'active tissues', we refer to foods such as fruits and vegetables after harvest or meat after dressing, in view of their biochemical activity. The principal biochemical post-harvest process that occurs in fruits and vegetables is respiration, whereby sugars are 'burnt ', oxygen is consumed and carbon dioxide is evolved. The rate of respiration is usually determined by measuring the rate of oxygen consumption or carbon dioxide release. Respiration is the most important (but not the only) cause of deterioration of fruits and vegetables during storage. It is usually said that the shelf life of fresh produce is inversely related to the rate of respiration. The rate of respiration is closely related to the temperature. The rate of respiration increases by 2 to 4 fold for every 10 °C increase in the temperature within the range of usual storage temperatures. However, too low storage temperatures may cause a condition known as 'chill injury' in certain commodities and must be avoided. Fruits and vegetables differ in the intensity of their post-harvest respiration rate. Following is a rough classification of some fresh produce with respect to their respiration rate during storage:

- High respiration rate: Avocado, asparagus, cauliflower, berries.
- Medium respiration rate: Banana, apricot, plums, carrot, cabbage, tomatoes.
- Low respiration rate: Citrus, apples, grapes, potatoes.

Many crops undergo a process of post-harvest ripening during storage. These are called 'climacteric' species, because their rate of respiration increases to a maximum in the process. Ethylene is produced and acts as a ripening hormone. Apples, bananas and avocado are climacteric, while citrus and grapes are non-climacteric. It is possible to control the rate of respiration of fruits, vegetables and cut flowers with the help of refrigeration alone or through the combination of refrigeration and 'controlled atmosphere' storage. 'Controlled atmosphere' consists of long-range refrigerated storage in closed chambers where the atmosphere has been artificially modified so as to depress the rate of respiration to the desired level. The optimum composition of the atmosphere depends on the commodity and often on the particular variety of the commodity. Table shows typical optimal storage conditions for selected fruits and vegetables.

Table: Optimal storage conditions for some fruits and vegetables.

Fruits in controlled atmosphere			
Produce	Temperature	Atmosphere	
Pineapple	10-15	% oxygen	%CO_2
Avocado	12-15	5	10
Avocado	10-15	2-5	3-10
Pomelo	10-15	3-10	5-10
Lemon	10-15	5	0-10
Mango	10-15	5	5
Papaya	10-15	5	10
Melon (cantaloupe)	5-10	3-5	10-15
Vegetables in ordinary cold storage			
Produce	Temperature (°C)	Relative humidity (%)	
Artichoke	0-2	90-95	
Asparagus	0-2	95-100	
Broccoli	0-2	95-100	
Carrot	0-2	98-100	
Eggplant	10-14	90-95	
Onion	0-2	65-75	
Potato	8-12	90-95	

Respiration is an exothermic process. The heat released by respiring commodities during refrigerated storage and transportation must be taken into account in calculating the refrigeration load required. Table gives the approximate rate of heat evolution during storage of selected commodities. Low temperature alone is not always sufficient for extending the shelf life of fruits and vegetables. Another important condition that must be controlled in post-harvest storage is relative humidity. Loss of water is often the cause of texture deterioration, wilting, shriveling etc. Water loss may be minimized by storing at high relative humidity, but humidity levels that can promote growth of mold must be avoided.

Table: Approximate rate of heat evolution during refrigerated storage of selected commodities, according to storage temperature.

Commodity	Heat evolution rate, Watts per ton, when stored at:			
	0 °C	5 °C	10 °C	15 °C
Apple	10-12	15-21	41-61	41-92
Cabbage	12-40	28-63	36-86	66-169
Carrot	46	58	93	117
Sweet corn	125	230	331	482
Green peas	90-138	163-226	–	529-599
Orange	9	14-19	35-40	38-67
Strawberry	36-52	48-98	145-280	210-275

The Effect of Low Temperature on Physical Properties

Numerous changes in physical properties caused by exposure to low temperatures may have significant effects on food quality. Some of these changes are:

- Increase in viscosity.
- Decrease in solubility, resulting in crystallization, precipitation, cloudiness (e.g.in beer).
- Hardening, transition to rubbery and glassy state in carbohydrate systems.
- Hardening of fats.
- Decomposition of colloidal systems such as emulsions and gels.

Freezing

Freezing is one of the most widespread industrial methods of food preservation today. The transition from chilling to freezing is not merely a continuous change that can be explained on grounds of the lower temperature alone. On the contrary, freezing represents a point of sharp discontinuity in the relationship between temperature and the stability and sensory properties of foods.

- The exceptional efficiency of freezing as a method of food preservation is, to a large extent, due to the depression of water activity. Indeed, when food is frozen, water separates as ice crystals and the remaining non-frozen portion becomes more concentrated in solutes. This 'freeze concentration' effect results in the depression of water activity. In this respect, freezing can be compared to concentration and drying.

- On the other hand, the same phenomenon of 'freeze concentration' may accelerate reactions, inducing irreversible changes such as protein denaturation, accelerated oxidation of lipids and destruction of the colloidal structure (gels, emulsions) of the food.
- The rate of freezing has an important effect on the quality of frozen foods. Physical changes, such as the formation of large ice crystals with sharp edges, expansion, disruption of the osmotic equilibrium between the cells and their surroundings, may induce irreversible damage to the texture of vegetables, fruits and muscle foods. It has been established that such damage is minimized in the case of quick freezing.

Effect of Freezing and Frozen Storage on Product Quality

For a very large number of food products, freezing represents the best preservation method with respect to food quality. The nutritional value, the flavor and color of foods are affected very slightly, if at all, by the process of freezing itself. The main quality factor that may be adversely affected by freezing is the texture. On the other hand, unless appropriate measures are taken, the deleterious effect of long-term frozen storage and of thawing on every aspect of product quality may be significant.

Effect of Freezing on Texture

In a vegetal or animal tissue, the cells are surrounded by a medium known as the extracellular fluid. The extracellular fluid is less concentrated than the protoplasm inside the cells. The concentration difference results in a difference in osmotic pressure which is compensated by the tension of the cell wall. This phenomenon, known as turgor, is the reason for solid appearance of meat and the crispiness of fruits and vegetables. When heat is removed from the food in the course of freezing, the extracellular fluid, being less concentrated, starts to freeze first. Its concentration rises and the osmotic balance is disrupted. Fluid flows from the cell to the extracellular space. Turgor is lost and the tissue is softened. When the food is thawed, the liquid that was lost to the extracellular space is not reabsorbed into the cell, but is released as free juice in the case of fruits or 'drip' in the case of meat.

It is generally believed that freezing damage to the texture of cellular foods is greatly reduced by accelerating the rate of freezing. This is explained by the fact that the condition of osmotic imbalance, created at the onset of freezing, disappears when the entire mass is frozen. Furthermore, rapid freezing results in the formation of smaller ice crystals, presumably less harmful to the texture of cellular systems. Another probable reason for the deterioration of the texture is the volumetric expansion caused by freezing. The specific volume of ice is 9% higher than that of pure water. Because cellular tissues are not homogeneous with respect to water content, parts of the tissue expand more than the others. This creates mechanical stress that may result in cracks. Obviously, this effect is particularly strong in foods with high water content such as cucumbers, lettuce and tomatoes. This kind of texture damage is partially prevented by adding solutes. Adding sugar to fruits and berries before freezing was a widespread practice before the development of ultra-rapid freezing methods.

The rate of freezing affects the size of ice crystals. Slow freezing produces large crystals. It has been suggested that large crystals with sharp edges may break cell walls and contribute to texture deterioration in cellular foods subjected to slow freezing. There is no controversy as to the fact that slow freezing results in higher percentage of drip in meat and fish. It is also accepted that rapid freezing causes less damage to the texture of particularly fragile fruits. On the other hand, the general applicability of the theory stating the superiority of rapid freezing seems to be questionable. Nevertheless, quick freezing continues to be the practical objective of food freezing process design.

Effect of Frozen Storage on Food Quality

Frozen storage, even at fairly low temperature, does not mean the absence of deteriorative processes. On the contrary, frozen foods may undergo profound quality changes during frozen storage. While the rate of reactions is generally (but not always) slower in frozen foods, the expected shelf life, and therefore the time available for the reactions to take place, is long. Some of the frequent types of deterioration in frozen foods are protein denaturation resulting in toughening of muscle foods, protein–lipid interaction, lipid oxidation and oxidative changes in general (e.g. loss of some vitamins and pigments).

Extensive studies on the effect of frozen storage on product quality were undertaken in the 1960s by researchers. A large number of commodities were tested for changes in chemical composition and sensory characteristics. A concept known as 'time temperature tolerance' (TTT) was developed. The studies showed a linear relationship between the storage temperature and the logarithm of storage time for equal reduction in quality (loss of a certain vitamin, loss of color or loss of organoleptic score). The logical conclusion of these studies could be that lower storage temperature always results in higher quality. This is not always the case, however. Reactions may be accelerated by the 'freeze concentration' effect more than they are attenuated by the lower temperature. In this case, the rate of deterioration (e.g. lipid oxidation) may increase as the storage temperature is lowered, passing through a maximum and then diminishing at very low temperatures. Mass transfer phenomena during frozen storage (oxygen transfer, loss of moisture) may be a major cause of quality loss. The quality of packaging is, therefore, particularly important in frozen foods. The PPP (product–process–package) approach consists of paying attention of all three factors in evaluating and predicting the effect of frozen storage on product quality.

Another type of change in frozen foods during storage is the process of recrystallization. Smaller crystals are more soluble than large ones. Equally, small ice crystals have a lower melting point then large ones. Consequently, if the storage temperature undergoes fluctuations, small ice crystals may melt and then solidify on the larger crystals. This may be the reason why foods frozen rapidly and those frozen slowly sometimes show similar ice crystal size distribution after storage. Recrystallization is particularly objectionable in ice cream, where conversion of small ice crystals to large ones results in loss of the smooth, creamy texture. The remedy is, of course, avoiding temperature fluctuation during storage as much as possible.

Freezing Methods

There are three major methods of freezing foods:

- Air Freezing:
    - Still-air 'sharp' freezer,
    - Air blast freezer,
    - Fluidized-bed freezer (IQF).
- Indirect Contact Freezing:
    - Single plate,
    - Double plate,
    - Pressure plate,
    - Slush freezer.
- Immersion and Cryogenic Freezing:
    - Heat exchange fluid,
    - Compressed gas,
    - Refrigerant spray.

Air Blast Freezing

Air freezing is the oldest and most common type of freezing used. The freezer section of household refrigerator and the deep freezer are examples of still air freezers or low air velocity systems.

Air blast freezing is a moderately fast freeze because of vigorous circulation of cold air. The product is placed on trays or mesh belts and passed slowly through an insulated tunnel. In different systems the temperature may range from -18 °C to - 34 °C, with an air velocity of 100-3500 lineal feet per minute, with a counter current air flow. Air blast freezers operate at lower temperatures than still air freezers and rely on movement of the cold air at high velocity over the food in order to achieve rapid removal of heat and to maximize the freezing rate.

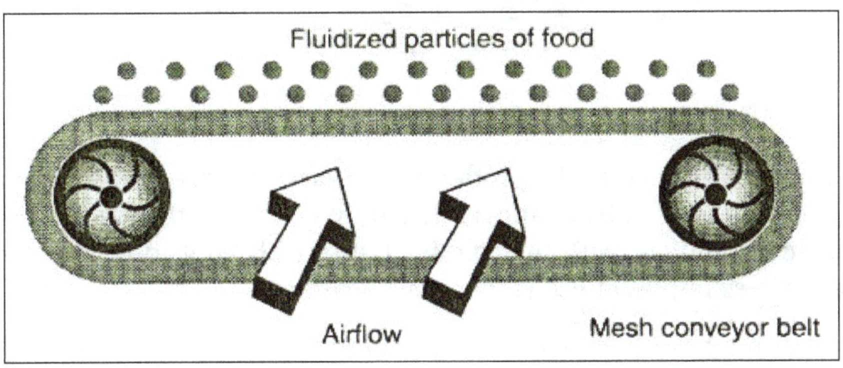

Fluidized-bed Freezer: In this type of air freezing, solid particles ranging in size from peas to strawberries are being exposed through a movement of the cold air (-18 °C to -34 °C) at high velocity as they pass along a conveyor belt. This will impart a vibratory motion to food particles, accelerating the freezing rate. The cold air being forced upward through the bed lifts and suspends the food particles, thus fluidization occurs. In this way, a rapid freezing rate is accomplished and an IQF (individually quick frozen) product is produced. In other words, food items are frozen as individual pieces and are not stuck together.

Advantages and Disadvantages of Air Blast Freezing; Fluidize-bed Freezing (IQF)

Advantages:

- Economical.
- Can freeze various sizes and shapes of food.
- IQF has more efficient heat transfer, increased rate of freezing.

Disadvantages:

- Possible excess dehydration (freezer burn).
- Undesirable bulging of the packages (by expansion of the product) may occur.
- Non-uniform products cannot be fluidized (IQF) easily.

Indirect Contact Freezers

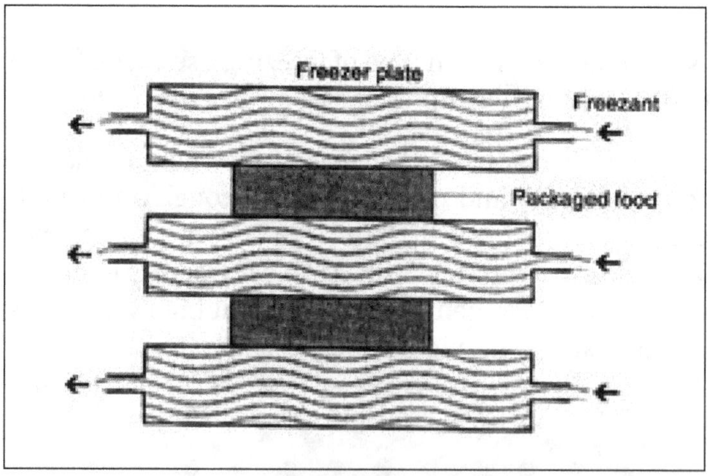

Plate Freezer.

Indirect Contact freezers are used in the production of various frozen food commodities. In these freezers, food is placed on belts or trays and a refrigerant circulates through a wall beside the food. As the food comes into "contact" with the cold wall, it quickly cools down and freezes. Plate and slush freezers are some examples of indirect contact freezers.

During plate freezing food products are placed in contact with a metal surface which is cooled by a cold brine, or a vaporizer refrigerant such as ammonia. The packaged food either rests on, slides against or is pressed between the cold metal plates. These plates maintain firm contact with two major surfaces of packages to facilitate heat transfer and prevent bulging of the packages during the freezing process. Fish sticks and frozen fish fillets are commonly frozen in plate contact freezers.

Another type of indirect contact freezer is the slush freezers or scraped surface heat exchangers. These freezers can be used only for fluid food products. A common example of a scraped surface freezer is the machine used to convert ice cream mix to soft ice cream in restaurants and ice cream shops. The same principle is used in the commercial production of ice cream that is sold as hard ice cream. In the case of ice cream, the rotator not only aids in promoting rapid freezing and the development of small ice crystals, but it also aids in the incorporation of air bubbles into the freezing mix which results in the formation of a solid foam.

Advantages and Disadvantages of Indirect Contact Freezing

Advantages:

- Economical,
- Minimal dehydration,
- Minimal package bulging.

Disadvantages:

- Slow freezing process,
- Products must be of uniform thickness.

Immersion and Cryogenic Freezing

Immersion freezing involves the immersion of packaged or un-packaged food products directly in a non-toxic refrigerant fluid. The refrigerant fluids commonly used are propylene glycol, glycerol, sodium chloride, calcium chloride, and mixtures of salt and sugar. Canned citrus juice, turkeys and chickens are often frozen in immersion freezing units. Ice cream popsicles can also be frozen using this method.

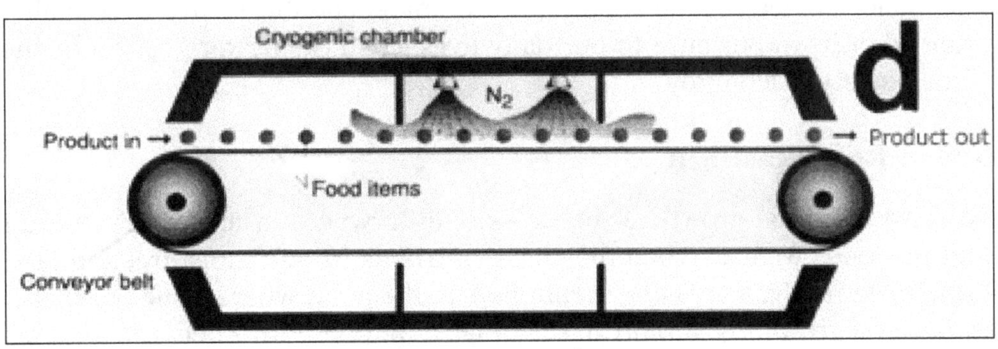

Cryogenic Freezing

Cryogenic freezing is accomplished with cryogenic liquids, with liquid nitrogen being the most commonly used. This is a very rapid freezing method in which un-packaged or thinly packaged foods are exposed to extremely cold freezant. In contrast to the liquid immersion freezing, heat removal is accomplished during a change of state by the freezant. Products such as TV dinners, preformed patties and other high value food products are frozen in cryogenic freezers because of the excellent retention of quality imparted by the rapid rate of freezing and small ice crystal formation. The figure on the right illustrates a cryogenic process.

Liquid nitrogen cryogenic freezing: The product is first placed on a conveyor belt and is moved into the pre-cooling part of the freezing unit. Once the food is cooled, the food is sprayed by liquid nitrogen as it is being moved through the conveyor belt; here is where the freezing process takes place, by the nitrogen boiling as it contacts the food. Finally, the food is allowed to equilibrate to the desired final temperature (between -18 °C to -30 °C).

Advantages and Disadvantages of Immersion/Cryogenic Freezing

Advantages:

- Rapid freezing process.
- Almost no dehydration.
- Oxygen is excluded, decreasing oxidative spoilage.
- Individual freezing pieces have less freezing damage.

Disadvantages:

- Difficult to find suitable freezants.
- Cost of operating is high.

Refrigeration

A refrigerator is one of the most important pieces of equipment in the kitchen for keeping foods safe. These electric units are so commonplace today, we forget a refrigerator was once little more than a box with a block of ice used to supply a rather undependable source of cold air. But we are instantly reminded of its importance to our daily lives when the power goes off or the unit fails, putting our food's safety in jeopardy.

Importance of Refrigeration

Refrigeration slows bacterial growth. Bacteria exist everywhere in nature. They are in the soil, air, water, and the foods we eat. When they have nutrients (food), moisture, and favorable temperatures, they grow rapidly, increasing in numbers to the point where some types of bacteria can cause illness. Bacteria grow most rapidly in the range of temperatures between 40 and 140 °F, the "Danger Zone," some doubling in number in as little as 20 minutes. A refrigerator set at 40 °F or below will protect most foods.

Types of Bacteria in Refrigerated Foods

There are two completely different families of bacteria: pathogenic bacteria, the kind that cause foodborne illness, and spoilage bacteria, the kind of bacteria that cause foods to deteriorate and develop unpleasant odors, tastes, and textures.

Pathogenic bacteria can grow rapidly in the "Danger Zone," the temperature range between 40 and 140 °F, but they do not generally affect the taste, smell, or appearance of a food. In other words, one cannot tell that a pathogen is present.

Spoilage bacteria can grow at low temperatures, such as in the refrigerator. Eventually they cause food to develop off or bad tastes and smells. Most people would not choose to eat spoiled food, but if they did, they probably would not get sick. It comes down to an issue of quality versus safety:

- Food that has been left too long on the counter may be dangerous to eat, but could look fine.
- Food that has been stored too long in the refrigerator or freezer may be of lessened quality, but most likely would not make anyone sick. (However, some bacteria such as Listeria monocytogenes thrive at cold temperatures, and if present, will multiply in the refrigerator over time and could cause illness.)

Safe Refrigerator Temperature

For safety, it is important to verify the temperature of the refrigerator. Refrigerators should be set to maintain a temperature of 40 °F or below. Some refrigerators have built-in thermometers to measure their internal temperature. For those refrigerators without this feature, keep an appliance thermometer in the refrigerator to monitor the temperature. This can be critical in the event of a power outage. When the power goes back on, if the refrigerator is still 40 °F, the food is safe. Foods held at temperatures above 40 °F for more than 2 hours should not be consumed. Appliance thermometers are specifically designed to provide accuracy at cold temperatures. Be sure refrigerator/freezer doors are closed tightly at all times. Don't open refrigerator/freezer doors more often than necessary and close them as soon as possible.

Safe Handling of Foods for Refrigerating

Hot food can be placed directly in the refrigerator or it can be rapidly chilled in an ice or cold water bath before refrigerating. Cover foods to retain moisture and prevent them from picking up odors from other foods. A large pot of food like soup or stew should be divided into small portions and put in shallow containers before being refrigerated. A large cut of meat or whole poultry should be divided into smaller pieces or placed in shallow containers before refrigerating.

Placement of Foods

The temperature in a refrigerator should be 40 °F or below throughout the unit, so that any place is safe for storage of any food. Raw meat, poultry, and seafood should be in a sealed container or wrapped securely to prevent raw juices from contaminating other foods.

Some refrigerators have special features such as adjustable shelves, door bins, crispers, and meat/cheese drawers. These features are designed to make storage of foods more convenient and to provide an optimal storage environment for fruits, vegetables, meats, poultry, and cheese.

Shelves

Shelves should be adjustable to accommodate a variety of packages. Tempered glass shelves are attractive and easy to clean. Some refrigerators feature sealed glass shelves to contain spills and make cleanup easier. Some shelves pull out to provide better accessibility to items in the back.

Specialized Compartments

Sealed crisper drawers provide an optimal storage environment for fruits and vegetables. Vegetables require higher humidity conditions while fruits require lower humidity conditions. Some crispers are equipped with controls to allow the consumer to customize each drawer's humidity level. An adjustable temperature meat drawer maximizes the storage time of meats and cheeses. Additional cool air is directed into the drawer to keep items very cold without freezing.

Safety of Foods Stored on the Door

Don't store perishable foods in the door. Eggs should be stored in the carton on a shelf. The temperature of the storage bins in the door fluctuate more than the temperature in the cabinet. Keep the door closed as much as possible.

Food Safety while Manually Defrosting a Refrigerator-freezer

Most refrigerators-freezers sold today don't require defrosting by the consumer. However, there are still units on the market and in homes that do allow frost to build up and require periodic defrosting. When food is removed from the freezer for defrosting and the unit is turned off, it's important to keep refrigerated foods cold and frozen foods from thawing. To do this, place the food in a cooler with a cold source or pack it in a box and cover it with blankets for insulation. Do not use any type of electrical heating device, ice pick, knife, or other sharp object to remove frost, as this could damage the inner lining.

Keeping the Refrigerator Clean

One very important step in keeping your food safe is keeping your refrigerator clean. Wipe up spills immediately - clean surfaces thoroughly with hot, soapy water; then rinse.

Once a week, make it a habit to throw out perishable foods that should no longer be eaten. A general rule of thumb for refrigerator storage for cooked leftovers is 4 days; raw poultry and ground meats, 1 to 2 days. Refer to the cold storage chart for storage of meat, poultry, and egg products in the home refrigerator.

To keep the refrigerator smelling fresh and help eliminate odors, place an opened box of baking soda on a shelf. Avoid using solvent cleaning agents, abrasives, and all cleansers that may impart a chemical taste to food or ice cubes, or cause damage to the interior finish of your refrigerator. Follow the manufacturer's instructions.

The exterior may be cleaned with a soft cloth and mild liquid dishwashing detergent as well as cleansers and polishes that are made for appliance use. The front grill should be kept free of dust and lint to permit free air flow to the condenser. Several times a year the condenser coil should be cleaned with a brush or vacuum cleaner to remove dirt, lint, or other accumulations. This will ensure efficiency and top performance.

Removing Odors

If food has spoiled in a refrigerator - such as during a power outage - and odors from the food remain, they can be difficult to remove. The following procedures may have to be repeated:

- Wipe inside of unit with equal parts vinegar and water. Vinegar provides acid which destroys mildew.
- Wash inside of unit with a solution of baking soda and water. Be sure to scrub the gaskets, shelves, sides, and door. Allow to air out several days.
- Stuff unit with rolled newspapers. Close the door and leave for several days. Remove paper and clean with vinegar and water.
- Sprinkle fresh coffee grounds or baking soda loosely in the bottom of the unit, or place them in an open container.
- Place a cotton swab soaked with vanilla inside freezer. Close door for 24 hours. Check for odors.
- Use a commercial product available at hardware and housewares stores. Follow the manufacturers' instructions.

Food Drying and Dehydration

Drying is an important and traditional process to remove the moisture from the food. The basic principle reason is that micro-organisms such as bacteria, fungi, mold required water for their growth and multiplication, which causes food spoilage and decay. Since water as a potential vehicle for pathogens in the food chain and it has to be removed to increase the shelf life of the food products. Drying and dehydration is an ideal process applicable to all food materials such as fruits, vegetable, cereals, pulses, milk, meat, fish etc to remove the moisture content. The moisture is an important factor in agricultural and food materials affects the shelf life of the product which due to microbial spoilage, oxidation and physical structure of the foods.

Free and Bound Water

Water is a basic element in foods which contains positively charged two hydrogen atoms and one negatively charged oxygen atom hence water is polar solvent. The bonds between oxygen and each hydrogen atom are polar bonds, having 40% partial ionic character. Water is the fundamental component in all living things and almost in all food materials contains more than 75% of water and fruits and vegetables contain water up to 95%. Water that can be easily removed from foods is known as free water, whereas water that cannot be removed easily is termed as bound water.

Several properties of foods significantly depend on water content. Solute molecules and ions, or chemical groups and surfaces, can be hydrated. There are several relations between water content and some macroscopic property because the part of water is present in the constituents. This water is called bound water. Hydration of water in food components involves small quantities of water. Only polar groups, and to a lesser extent dipoles, can "bind" water molecules.

Free Moisture

Free moisture is the liquid in excess of the equilibrium moisture content at a given temperature and humidity. Free moisture can be easily removed by drying by vapor pressure difference between the product and ambient conditions.

Bound Moisture

Bound moisture in a food material is that liquid which is retained in a manner that exerts a vapor pressure less than that of the free liquid at the given temperature.

Liquid may become bound to the solidby retention in small capillaries, by solution in cells or fiber walls, by homogeneous solution throughout the solid or by chemical or physical adsorption on solid surfaces.

Many food constituents can hold onto water molecules, this water cannot be removed simply and they do not behave like free water. Some properties of bound water are:

- Bound water cannot act as a solvent for salt, minerals and sugars present in food material.
- Freezing point of bound water is lower than the free water.
- Bound water exhibits essentially no vapor pressure.
- The density of bound water is greater than that of free water.

Unbound Moisture

Unbound moisture exerts a vapor pressure equivalent to that of the free liquid. Thus in a non-hygroscopic material, all the liquid is unbound and in a hygroscopic material, the unbound moisture is the liquid in excess of the equilibrium moisture content (EMC), corresponding to saturation humidity.

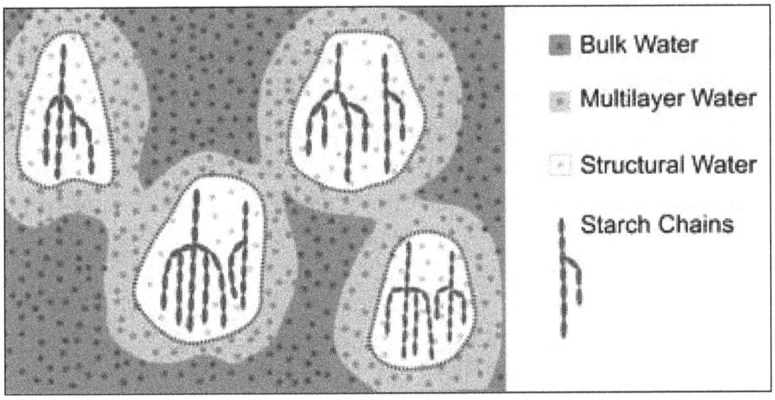

Moisture profile in food.

Water molecules attach to ionic sites on molecules like starches, pectin and protein. Water adjoining to these molecules has firm bonds between them; the following water layers are attached less firmly, finally the structure of free water exists. Water may be entrapped in foods such as gels, fruits and vegetables. Entrapped water is not easily mobilized in cells, but if released during physical damage, it flows freely.

Water Activity

The availability of water for biological reactions and support the growth of micro-organisms is termed as water activity. Water activity exhibits the ability of type of microorganisms to grow in food materials. If water activity decreases, the ability of microorganism to grow will also decrease. Water activity (a_w) is expressed as the ratio of the vapor pressure in a food (P) to the vapor pressure of pure water (P_o).

$$a_w = P / P_o$$

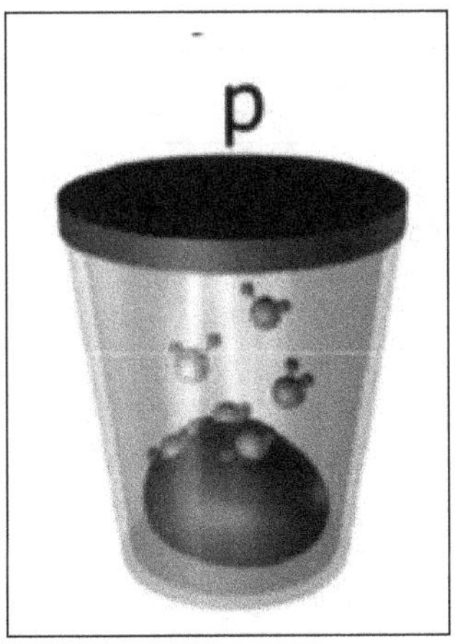

Food material

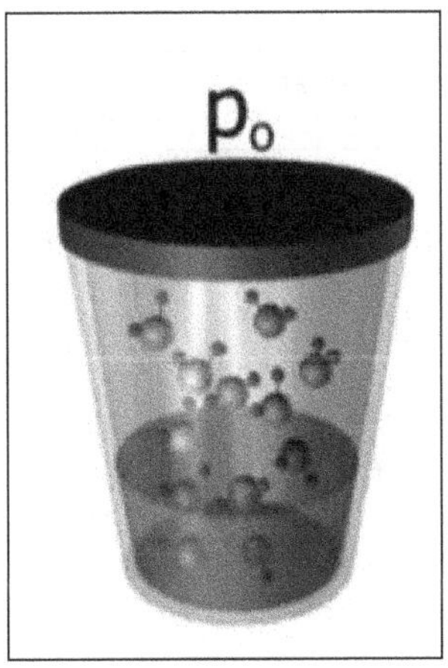

Pure water

The a_w is directly related to the microbial, chemical, and physical properties of foods. The range of water activity is closely associated with changes incolor, aroma, flavor, texture, stability, and acceptability of fresh and processed food products. Table depicts the water activity of selected food products. Figure shows the inhibition the growth of micro-organisms in foods, The gram negative bacteria can grow water activity ranges from 0.87 -0.91, where the Xerophilic microorganism can grow and multiply in extreme dry conditions in the water activity range between 0.6-0.65. Dry fruits, dehydrated fruits and vegetables, spices and nuts are targeted by this microorganism. Osmophilic organisms are microorganisms adapted to environments with high osmotic pressures, such as high sugar concentrations and the water activity of 0.62 and Halophiles are organisms that thrive in high salt concentration.

Table: Water activity of selected foods.

Food products	Water activity
Fresh meat, poultry, fish	0.99 - 1.00
Fresh fruits, vegetables	0.97 - 1.00
Eggs	0.97
Natural cheeses	0.95 - 1.00
Baked cake	0.90 - 0.94
Uncooked rice	0.80 - 0.87
Cured meat	0.87 - 0.95
Jellies	0.82 - 0.94
Sweetened condensed milk	0.83
Jam	0.75 - 0.80
Honey	0.75
Dried fruit	0.55 - 0.80
Dried whole egg	0.40
Dried whole milk	0.20
Cereal	0.10 - 0.20
Sugar	0.19

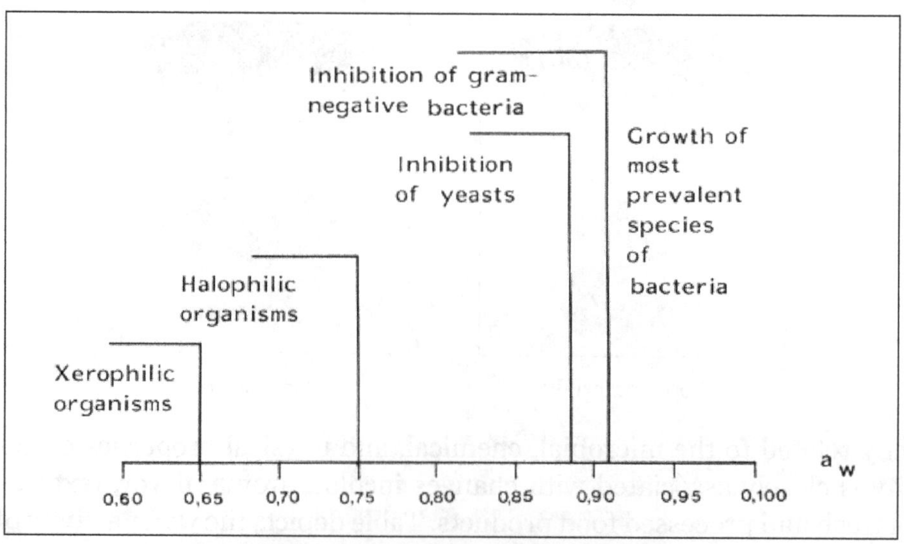

Inhibition water activity of microorganism.

Relationship between Water Content and Water Activity

Water content on its own is not enough information to determine food safety or predict product shelf life. The relationship between water content and water activity is complex and related to the relative humidity of the food and its water content. This relationship must be determined for each specific food item.It is easy to assume that foods with higher water content will have a higher water activity than dry foods. It is possible to have products with the same water content but very different water activities. For example, salami and cooked beef have similar water content of approximately 60 per cent. However, the water activity of salami is 0.82 and cooked beef is approximately 0.98.

Adsorption and Desorption Characteristics of Water

All food materials are hygroscopic in nature and exhibits different equilibrium moisture characteristics depending upon whether the equilibrium is reached by adsorption or desorption of the moisture. This phenomenon is known as hysteresis.

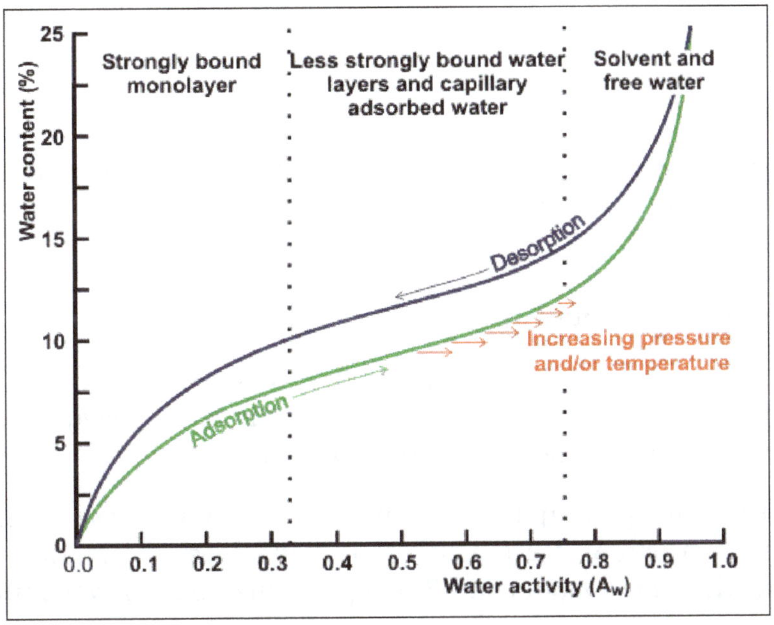

Adsorption and desorption characteristics of food material.

Properties of Water

Properties of water are important for drying and dehydration of food materials. The properties such as specific heat, latent heat, latent heat of fusion, latent heat of vaporization, sublimation, evaporation etc are essential while studying drying characteristics of foods.

Specific Heat of Water

The energy required to raise the temperature of 1 g of water by 10C is termed as specific heat of water. Due to the presence of hydrogen bonds the specific heat of water is relatively high when compared to other substances. The specific heat of water is 1 cal/g / °C. which means, it takes 100 cal to raise the temperature of 1 g of water from 0 to 100 °C. Once ice has reached 0 °C, energy needs to be put in to break the hydrogen bonds and enable ice to change to the liquid form. Until the ice has been converted to liquid, there is no further change in temperature until steam is created.

Latent Heat

When something absorbs heat without changing its temperature, the heat absorbed is called latent heat. Thus each gram of water at 100° C at ordinary pressures can absorb up to 540 calories without changing its temperature as the heat absorbed is used to break the hydrogen bonds among the water molecules and not to change the temperature. This means that 1 gram of water at 100° C changes to 1 gram of steam at 100 °C by absorbing 540 calories.

- Latent Heat of Fusion: The energy required to convert 1 g of ice to water at 0 °C and is 80 cal; that is, 1 g of ice at the freezing point absorbsapproximately 80 cal as it changes to the liquid state.
- Latent heat of Vaporization: The energy required to convert 1 g of water into vapor at 100 °C and is 540 cal; that is, 1 g of water at the boiling pointabsorbs approximately 540 cal as it becomes steam.

Sublimation

The conversion of icecrystals to vapour without going through the liquid phase, when the heat is applied under vacuum condition is called as sublimation. Sublimation is important in freeze drying of food materials.

Vapor Pressure and Evaporation

The process of evaporation in a closed container will proceed until there are as many molecules returning to the liquid as there are escaping. At this point the vapor is said to be saturated, and the pressure of that vapor is called the saturated vapor pressure.

At high vapor pressure, the liquid vaporized easily and many molecules exist in the vapor state at low boiling point. The lowvapor pressure signifies the liquid does not vaporize easily and that there are few molecules existing in the vapor state. The liquid boils when the vapor pressure reaches the external pressure.

Vapor pressure increases with increase in temperature. The molecules have more energy at higher temperature and the molecules are easily overcome the forces holding them within the liquid and to vaporize; hence the range of molecules are higher in the vapour state. Salt and sugar molecules dilute the water and very few molecules are available for vaporization and thus, there will be fewer molecules in the vapor state and the vapor pressure will be lower.

Drying and Dehydration Process

In drying mechanism of water removal involves two concurrent processes, namely, transfer of heat for the evaporation of water to the food and transport of the water vapors formed away from the food. Drying is, therefore, an operation based on simultaneous heat and mass transfer.Drying and dehydration occurs by three modes viz., heat, dry air and air movement. The air in the drying chamber should hot enough with low relative humidity and good air velocity for proper drying. The food product should dry until it reaches the equilibrium moisture content.

Principle of Drying and Dehydration Process

Drying is the process of moisture removal from the product, or grain in this case. Since foods are hydroscopic material which can either absorb or desorb moisture from the air or its surroundings depending on the difference in vapour pressure, moisture transferred from a higher vapour pressure, to the lower one.

In the sundrying process, food is heated by solar radiation thus creating a higher vapour pressure in grain than the surrounding air. In the same manner, the heated air drying process starts when the grain is heated (by conduction) when it comes in contact with the air. Higher velocity air flow in heated air drying has the advantage of reducing the boundary layer of the grain, thereby increasing the heat transfer coefficient of the grain, as well as increasing the rate of moisture movement from grain to the surrounding air. Therefore, the drying rate of a specific kind of food material is dependent on both air temperature and air flow rate.

Importance of Drying

- Enhances long time storage of foods without deterioration.
- Enhances continuous supply of product throughout the year.
- Reduces the post-harvest losses and increase the utilization of food materials even in off seasons.
- Reconstitution of dried material by addition water is possible.
- Enhances the value addition of food materials and hence the quality of the food product is improved.

Measurement of Moisture Content

The measurement of the total water present in the food material is known as water content, or moisture content. Moisture content is measured in terms of dry basis and wet basis. Wet basis moisture content is described by the percentage equivalent of the ratio of the weight of water (W_w) to the sum of weight of material and the weight of the dry matter (W_d).

$$Mw(\text{wet basis}) = \frac{W_m}{W_w + W_d} \times 100$$

Wet basis moisture content ranges from 0 to 100 per cent. Wet basis moisture is used to describe the water content of agricultural materials and food products.

Dry basis moisture content is described by the percentage equivalent of the ratio of the weight of water (W_W) to the weight of the dry matter (W_d). Dry weight moisture content ranges from 0 to very large percentages. Dry basis moisture is most commonly used for describing moisture changes during drying. When a sample loses or gains moisture, the change in the dry basis moisture is linearly related to the weight loss or gain.

$$M_d\left(\text{dry basis}\right)=\frac{W_w}{W_d}\times\left(100\right)$$

Equilibrium Moisture Content

When a solid is exposed to continual supply of air at constant temperature and humidity, have a fixed partial vapour pressure of the vapour. The solid either lose moisture by evaporation or gain moisture from the air until the vapour pressure of the moisture of the solid equals to the vapour pressure of the air. The solid and the air are then in equilibrium, and the moisture content of the solid in equilibrium with the surrounding conditions is known as equilibrium moisture content (EMC). The EMC is used to determine the absorption and desorption characteristics of food material under a specific temperature and relative humidity conditions. Since different materials have different EMC and thus EMC is directly related to drying and storage.

Constant Rate Period

In the initial period of drying, the behavior of food materials follows constant rate period and followed by falling rate period. In constant rate period, the rate of evaporation under any given set of air conditions is independent of the solid and is essentially the same as the rate of evaporation from a free liquid surface under the same conditions. The rate of drying during constant rate period is based on the:

- Difference between the temperature of air and temperature of the wetted surface at constant air velocity and relative humidity.
- Difference in humidity between air stream and wet surface at constant air velocity and temperature.
- Air velocity at constant air temperature and humidity.

In constant rate period drying takes place by surface evaporation and moisture moves by vapour pressure difference. The moisture content at which the drying rate ceases to be constant is known as critical moisture content. The critical moisture constant of a product also depends upon the characteristics of a food material such as shape, size and drying conditions.

Falling Rate Period

Cereal grains are usually dried entirely under falling rate period. The falling rate period enters after the constant drying rate period and corresponds to the drying cycle where all surface is no longer wetted and wetted surface continuously decreases until at the end of this period. The cause of falling off in the rate of drying is due to the inability of the moisture to be conveyed from the centre of the body to the surface at a rate comparable with the moisture evaporation from its surface to the surroundings.

The falling rate period is characterized by gradual increase in temperature both at the surface and within the food materials. The changes in the air velocity have much smaller effect during the constant rate period. The movement of moisture within the food material ie., from the centre to the surface by liquid diffusion and the removal of water content from the surface of the food material to the atmosphere are the two important controlling factor in drying of foods. The falling rate period consists of two stages:

- First Falling Rate: The moisture is transported from the inside of the product to the surface and the critical moisture content is reached. However, dry spots appear on the surface and surface area evaporation decreases.
- Second Falling Rate: In the second falling rate period, there is a slow diffusion of water to the inner surface leading to desorption and diffusion through pores to the surface. In conventional heat treatment, products are dried at higher temperature and for a longer drying time, creating thermal damage.

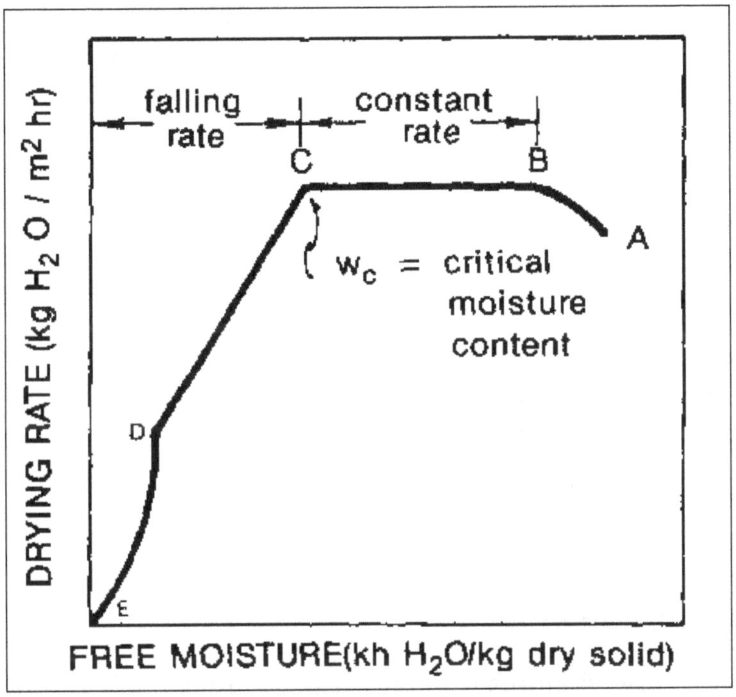

Drying characteristics of food material

The mechanism of drying process is as follows:

- Liquid diffusion surface: Water movement due to moisture concentration difference.
- Capillary theory of drying: Water movement due to surface forces.
- Surface diffusion: Water movement due to moisture diffusion in the pores present in the food materials.
- Vapour diffusion: Vapour movement due to difference in vapour pressure.
- Thermal diffusion: Vapour movement due to temperature difference.
- Hydrodynamic flow: Water and vapour movement due to total pressure differences.

Canning

The process of sealing foodstuffs hermetically in containers(of metal, glass, thermo stable plastic, or a multi-layered flexible pouch) and sterilizating them by heat for long storage is known as canning. It enhance the shelf life of the products ranging from 1to5years, although under specific conditions it can be much longer. Heating is the principle factor to demolish the microorganisms and the everlasting sealing is to prevent re-infection.

Importance of Canning

The fresh foods are highly perishable and spoil/lose their quality due to high levels of water content. Microorganisms can grow or multiply rapidly on the surface of foods. The parameters such as moisture, oxygen and enzymes are present throughout fresh food tissues, that influences the growth of microorganism that spoil the food stuffs. To minimize the effect of microorganism proper canning practice to be followed. Main purpose of this method is to preserve the foods by thermal treatment so that it can be safely consumed at a later time. Consumer safety is the primary goal when food is canned.

Nutritional Value of Canned Food

Many vegetables begin to lose vitamins when harvested. Almost half of the vitamins may be lost within a few days if not the fresh produce is cooled or preserved. Within one to two Refrigerated produce may lose half of its vitamins within one or two weeks. Canning destroys from 1/3 to 1/2 of vitamins A and C, thiamin, and riboflavin when subjected to heat treatment. Each year on storage the canned foods accounts for a loss of sensitive vitamins from 5 to 20 percentage. The amounts of other vitamins are slightly lower in canned foods compared with fresh foods. Vegetables that are handled properly and canned quickly after harvest may be more nutritious than fresh produce held many days after harvest. The heating process in canning appears to make dietary fiber more soluble and more readily fermented in the colon into gases and physiologically active by-products.

Canning Preservation of Foods

Fresh foods spoil or lose their quality for several reasons:

- Growth of undesirable microorganisms-bacteria,

- Molds and yeasts,
- Activity of food enzymes,
- Reactions with oxygen,
- Moisture loss,
- Destroys microorganisms,
- Inactivates enzymes,
- Containers are sealed to avoid recontamination.

Containers for Canned Foods

The container plays essential role in food canning, it must be:

- Capable of being hermetically sealed to avoid access of microorganisms.
- Resistant to liquids and gases, including water vapour.
- Retain the condition of biological stability (i.e, commercial sterility) that was induced by thermal treatment alone or combined with other chemical and physical process.
- It defend the contents against damage during transportation and storage.

Can Materials

Broad varieties of materials are used for manufacture of cans for preservation of foods and are mostly packed in metal containers.

Steel

Materials commonly used to make metal food cans are tinplate, tin-free steel, and nickelplated steel. The amount of tin used is about 1.5% of the can's weight and be supposed to contain less than 1% lead and is commonly used to prevent rusting. To avoid contact between foods and metal, organic material are usually coated on the inside of the can. The commonly used organic coatings in food industry are:

- Acid-resistant
- Sulfur-resistant

Acid-resistant coated cans are primarily used for fruit products whereas meat products are usually packed in cans that have been lined with sulfur-resistant materials.

Glass Containers and Metal Closures

Although the ample range of glass containers and metal closures for canned foods are used. It has a high conductivity of heat and cannot be broken easily. Being dense, any probable bad effects of light on food stuffs are avoided. It can endure the stresses forced during thermal processing and cooling.

Important Food Groups

Group	pH	Foods
Low acid	5.0 to 6.8.	Dairy products, vegetables, meats, seafood and poultry
Acid food	4.5 and 3.7.	Fruits for example oranges, pear, tomato and apricots
High acid	3.7 to 2.3	Fermented foods, Jam, Jellies, pickles

Packing the Jar Plays a large role in heat transfer through the product Can be completed by using one of the following packing processes: raw or cold pack hot pack.

Raw or Cold Pack

Places raw food directly in the jar and then hot, boiling liquid is poured over the contents.

Hot Pack

Involves cooking foods in liquid before packing and then the cooking liquid is poured over the food in the jar.

Headspace

It is the area in the jar between the inside of the lid and the top of the food or its liquid Is usually the following for certain foods:

- Jellies- ¼ inch fruits
- Tomatoes and pickles- ½ inch
- Low acid foods- 1 inch to 1 ¼ inches:
    - If too little, can result in the following: food bubbling out during processing proper seal may not occur because deposits can form on the rim.
    - If too much, can result in the following: food at the top of the jar can become discolored proper seal may not occur because not all of the air may be forced from the jar.

Canning Methods

Canning can be done by two methods boiling water bath method and the pressure canner method.

Water bath canning.

Pressure canning.

Boiling Water Canning

This is the most commonly used commercial method of canning for food containing high acid content and it requires shorter time, lower temperature for processing of foods. Place six inches of water in the canner for hot packed jars, the water should be simmering at 180 °F for raw or cold packed jars, the water should be heated to 140 °F and the jars are placed on a rack so that they do not touch the bottom of the canner. The water should rise to one to two inches above the tops of the jars, more hot or boiling water can be added if necessary. After processing time, canner is turned off, the lid is removed. Remove jars straight out of canner and let it cool for 12 to 24 hours. The seals are checked and the rings are removed. Jars are then stored in a cool, dry, dark place.

Advantages

It include easy for beginners to learn best method for fruits, high acid and pickled foods inexpensive to start.

Disadvantages

It include temperature does not get high enough to ensure the destruction of microorganisms in food without high sugar or acid content longer time period to complete canning process.

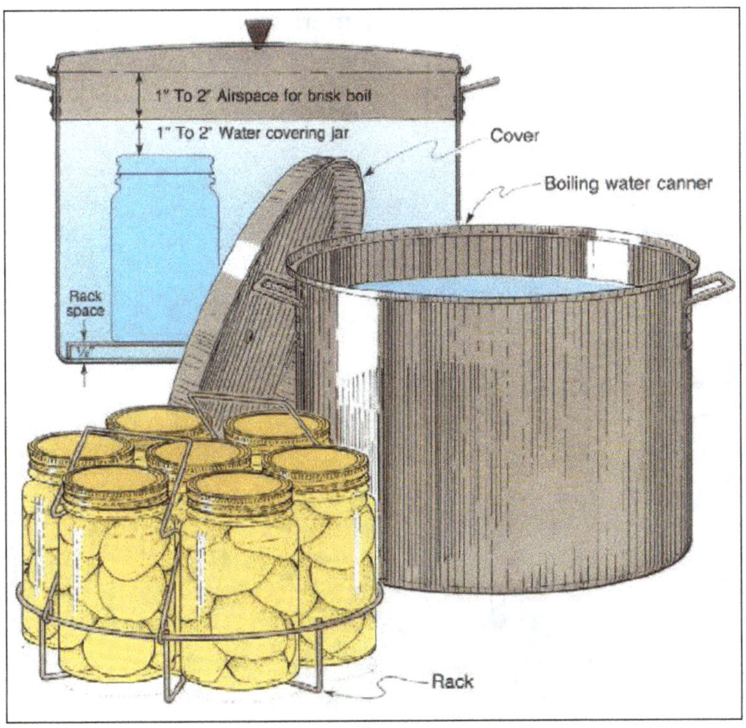

Pressure Canning

In this method the food products are exposed to 240° F, reduce the risk of food borne bacteria. Low acid foods such as vegetables, meats, poultry and seafood are canned by this method to remain foods fresh and safe to eat. Combination of high-acid and low-acid foods can be conserved for a longer period. A spore forming microorganism called Clostridium botulinum are killed at boiling temperatures, can withstand these temperatures. In low acid foods the spores will germinate in the absence of air, which forms the toxic botulinum. These spores can only be destroyed by pressure cooking method at a temperature of 240°F, or above, for a particular amount of time depending on the type of food and altitude. Clostridium botulinum spores cannot grow at high acidic conditions that is pH of 4.6 or less and this type of foods can be safely canned using the boiling water bath method. Foods that have a pH value near to 4.6 need to have acid added to them in order to use the water bath method. This is accomplished by adding lemon juice of citric acid. Place two to three inches of water simmering or hot in a canner and lid is placed on canner with weight off. Allow a steady stream of steam to escape for 10 minutes, close vent or petcock Count time when correct pressure is reached Turn off heat at end of processing time and let pressure drop to zero. Wait two minutes after pressure is at zero remove weight or petcock and wait ten minutes. Open canner and remove jars, cool the jars for 12 to 24 hours and then check that the jars are properly sealed.

Advantages of Pressure Canning

Include: best method for low acid foods, such as vegetables foods reach a higher temperature process is completed in a short period of time.

Disadvantages of Pressure Canning

Include: start-up cost is higher than boiling water method as a pressure cooker is required can be dangerous if pressure is not monitored and released some soft fruits can be damaged by high heat.

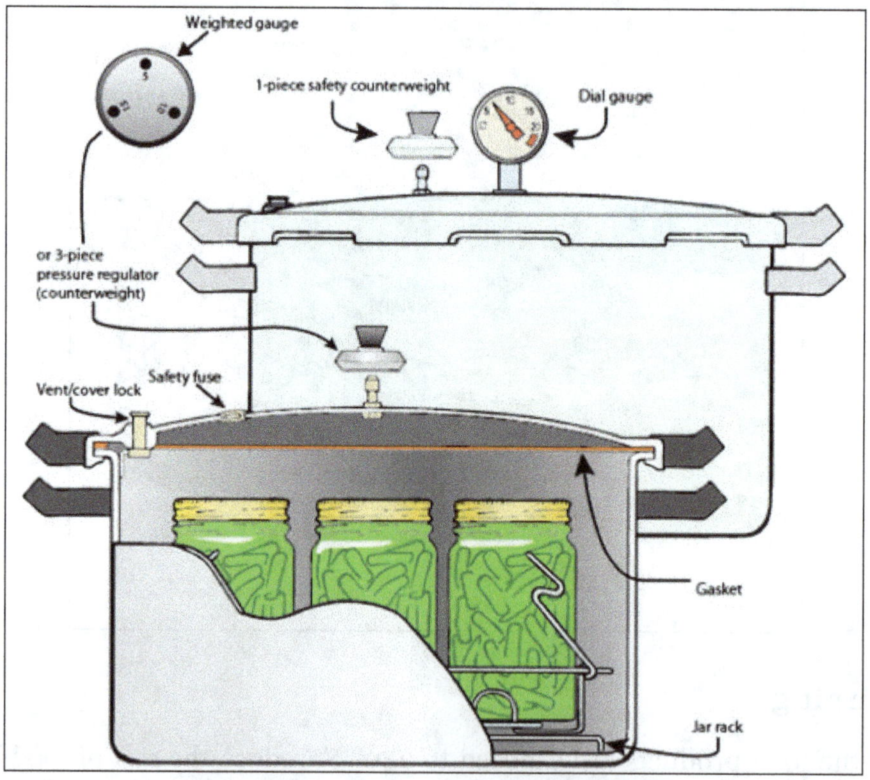

Canning time and temperature depends on:

- The physical state of the food.

- Size of Pieces.
- Fullness of pack.
- Container size and material.
- Initial temperature of food.
- The heat resistance of microorganisms or enzymes likely to be present in the food.

General Process Flow Chart of Canning

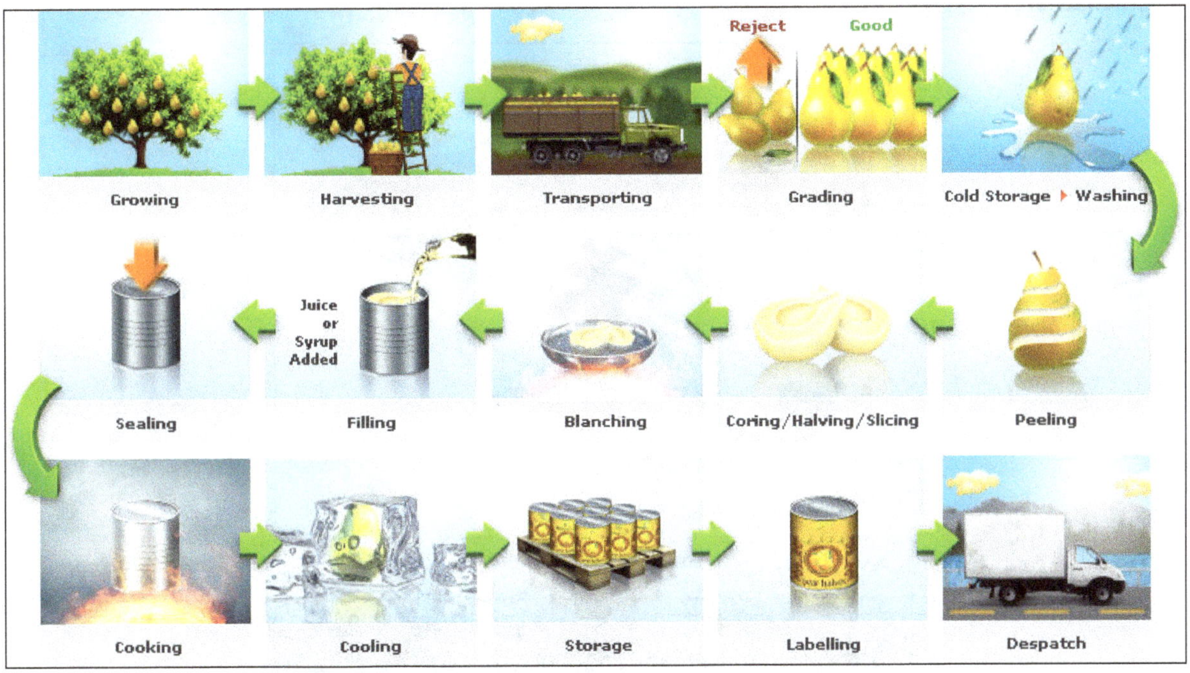

Processing

SELECTION
↓
SORTING/GRADING
↓
WASHING
↓
PEELING, CORING, PITTING
↓
BLANCHING
↓
CAN FILLING
↓
SYRUPING/BRINING
↓
LIDDING/CLINCHING
↓
EXHAUSTING
↓
SEALING
↓
WASHING
↓
COOLING
↓
LABELLING AND STORAGE

Selection

The fresh fruits and vegetables were used for canning. The selected fruit should be firm, mature and uniformly ripened. It should be free from all unpleasant blemishes, insect damage and malformation. Over-ripe fruit is normally infected with microorganisms and results in poor quality.

Sorting and Grading

After the preliminary sorting, the fruits and vegetables are graded. according to their size, color etc. It is usually done by manual or by mechanical grading (screen graders, roller graders, rope or cable graders etc).

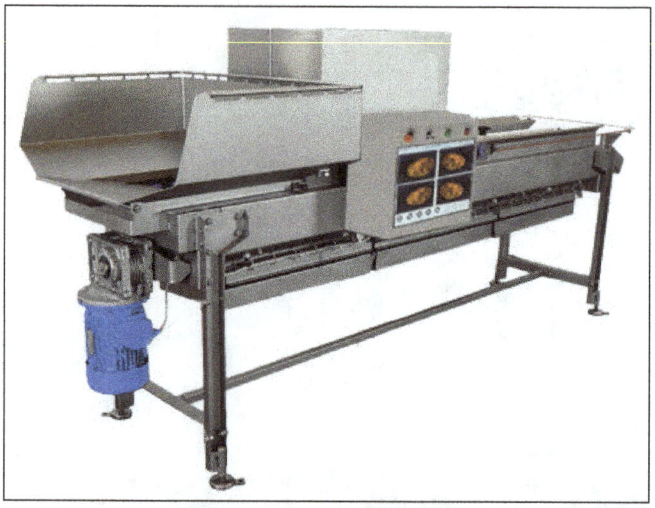

Mechanical Grader

Washing

The graded fruits and vegetables are washed to remove the foreign matter, dirt, dust with water in diverse ways such as soaking or agitation in water, washing with cold or hot water sprays, etc. Among all the methods spray washing is more efficient. Vegetables may first be soaked in potassium permanganate solution to disinfect them.

Peeling, Coring and Pitting

After washing the fruits and vegetables are peeled to remove the outer layer. There are 3 common methods of peeling:

- Mechanical peeler,
- Hand peeling by machine,
- Lye peeler(dipping the fruits and vegetables in a solution of boiling caustic soda or lye solution of strength 1-2%for 30 seconds to 2 minutes).

Cores and pits in fruits are removed by manual or by mechanical means.

Hand Peeling

Lye peeling

Mechanical peeler

Blanching

Treatment of fruits and vegetables with boiling water or steam for shorter period of time followed by cooling prior to canning, is called 'blanching'. It is commonly done to remove or loosen the outer covering(skin).This process can reduce the microbial growth and contamination. It helps to inactivate the reaction of enzymes which inturn prevent the discoloration in fruits and vegetables. It improves the flavor by reducing the astringency in food products.

Can Filling

The cans used for filling of food products are washed and exposed to steam jet which helps to get rid of any adhering dust or foreign matter in it. Prior to filling of contents(fruits and vegetables) a small amount of syrup(for fruits) or brine(for vegetables) is transfer to the cans which serves as a medium to contents. This filling operation can be done by machine or byhand filling.

Syruping and Brining

The hot sugar syrup for fruits of concentration of 35-40% and hot brine solution for vegetables of concentration 1-2% normally used to fill the cans. The syrup or brine should be added to the can at a temperature of 79 °C to 82 °C, leaving a headspace in the can so that when the filled can is closed on the double seaming machine, the headspace left inside ranges from 0.32 cm to 0.47 cm. This step is done to progress the taste of the canned product and to fill up the inter space between fruits and vegetables.

Lidding or Clinching

Cans after being filled, are covered loosely with lid and passed through the exhaust box. Now, lidding is replaced by clinching process where the lid is partly seamed to the can by a single first roller action of double seamer.

Exhausting

This operation reduces the risk of corrosion of tin plate and pin holing during the storage and staining of the product is minimized as the oxidation process is prevented. the exhausted cans are then passed into trough containing 82-87 °C of water or it can be sent to a steam box by a moving conveyor belt. The time taken for this process usually varies between 5-25 minutes depending on the nature of the substance. The spoilage of the canned products are prevented product by stopping the chemical reactions which results in bulging of can.

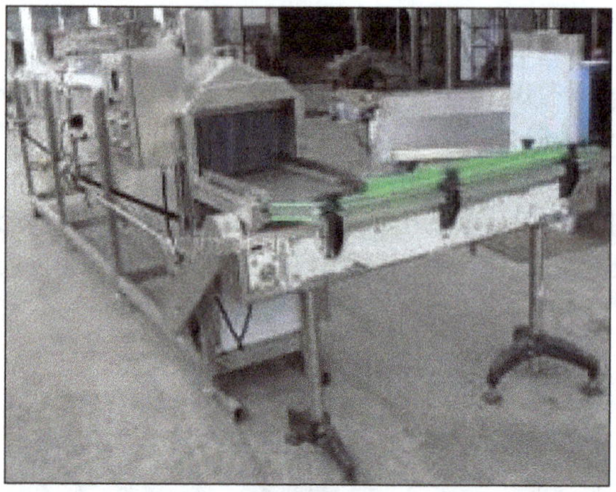

Sealing

After exhausting, special closing machines are used to seal the cans which are known as double seamers. It may be hand operated or semi-automatic/fully automatic seamers.

Processing/Sterilization

Sterilization is done to reduce the growth of microorganisms that causes spoilage. The temperature required for processing of fruits is 100 °C whereas for vegetables is 116-120 °C.The overall time required to sterilize canned food is mainly depends on the size of can, the processing temperature, the rate of heat penetration at the center of the can, pH of the food product, the type and number of organisms present in it.

Washing and Cooling

After the cans are closed, they pass through a detergent spray washer to get rid of grease and other material. The washing should consist of hot water (66 °C) then by appropriate pre-rinse by detergent spray wash. It is then followed by a fresh warm water rinse (66 °C).Instantly after processing, the cans are cooled in water to a temperature of 36 °C to 42 °C. to avoid thermophilic spoilage or can rust. If the cans are cooled much below 36 °C, they may not dry thoroughly and results in rusting. If the cans are cased at temperatures much over 42 °C, thermophilic spoilage may occur.

Labeling and Storage

The cans are labeled, packaged and stored at a clean and dry place. Storage temperatures of sterile canned meat products should not be above 21.1 °C, because higher temperatures obviously speed up deterioration during storage, thus limiting shelf life.

Effects of Canning on Foods

Colour

The time–temperature combinations used in canning have effect on naturally occurring pigments in foods. In fruits and vegetables, chlorophyll is converted to pheophytin, carotenoids carotenoids are isomerised isomerised from 5, 6-epoxides epoxides to less intensely intensely coloured 5, 8-epoxides, and anthocyanins are degraded to brown pigments. Discolouration of canned foods during storage occurs, for example, when iron or tin react with anthocyanins to form a purple pigment, or whencolourless leucoanthocyanins form pink anthocyani n complexes in some varieties of pears and quinces.

Flavour and Aroma

In canned meats there are complex changes, for example pyrolysis, deamination and decarboxylation of amino acids, degradation. In fruits and vegetables, changes are due to complex reactions which involve the degradation, recombination and volatilisation of aldehydes, ketones, sugars, lactones, amino acids and organic acids.

Texture or Viscosity

In canned meats, changes in texture are caused by coagulation and a loss of water holding capacity of proteins, which produces shrinkage and stiffening of proteins, which produces shrinkage and stiffening of muscle tissues. The texture of solid fruit and vegetable pieces is softer than the unprocessed food due to solubilisation of pectic materials but is considerably firmer than canned products.

Advantages of Canning

- It modify the foods by varying the moisture, pH, or salinity levels to shield against microbes, bacteria, mold, and yeast.
- The activity of enzymes are limited to certain levels.
- The shelf life of the canned foods can be kept for 1-2 years or longer.
- It is an excellent way to decrease the environmental impact.

Limitations of Canning Process

- Glass jars can be broken easily.
- Seals can be broken and it caused spoilage.
- It is a time consuming method.
- It does not taste as good as fresh foods.
- It also requires a major investment of time and equipment.

Spoilage of Canned Foods

Spoilage of canned foods according to the condition and content of the can:

- Swell: It is the bulging of both sides of can by positive internal pressure due to generation of gas by microorganisms or by chemical action. It may be hard or soft swell.
- Flipper: The appearance of the can will be normal but there may be flip at any one end, when the can is struck next to a solid object but snaps back to the ordinary condition under light pressure.
- Springer: It is a defect that a can bulged from one end which if forced back into normal position, the opposite end bulges.
- Overfilled can: It has convex ends due to overcapacity and not regarded as spoiled.

Spoilage of canned food according to the causes:

- Microbial Spoilage: This type of result from inadequate processing or leakage. There are 3 types of spore forming bacteria:
    - Gas producing aerobes and anaerobes with optimum growth temperature 37 °C.
    - Gas producing anaerobes with optimum temperature 55°C.
    - Non-gas producing aerobes or facultative anaerobes with optimum temperature 55°C generate flat souring bacteria. Leak can be detected by the bubbles when compress under water. There is a loss of vacuum (concavity) when heated to 38°C followed by slow cooling.
- Flat Souring: It results in high acid formation without gas production. It has a sour odour, bitter taste. It is caused by thermophilic bacteria such as,
    - Bacillus coagulans,
    - Bacillus stearothermophilus,
    - Bacillus circulans.

These bacteria attack carbohydrate producing acid without gas. It is common in tropical and sub-tropical countries. The affected cans should be destined.

- Chemical Spoilage:
    - Hydrogen Swell: Creation of hydrogen gas in can due to internal corrosion or scratch. It happen mainly in acidic foods (canned fruits).It is not related to fermentation or bacterial spoilage. There is a unreliable degree of bulging, when opened there is a release of odorless burnable gas. It is risk-free but undifferentiated from swell of spoiled can.
    - Sulphiding (Sulphur Stinker Spoilage): It leads to staining of inside of can with pink to dark purple. This happens due to the reaction of sulphur-containing proteins (liver, kidney, tongue) with liberated hydrogen sulphide from bacterial spoilage (Cl. nigrificans (sulphur stinker)) with the odor of rotted egg. It may be accompanied with blackening when hydrogen sulphide react with steel base of tin forming iron sulphide and it may direct to pitting. Sulphiding can be prohibited by sulphur-resistant lacquer.

Food Freezing

Freezing is the removal of heat from the packaged or whole foods resulting in the temperatures between slightly below the freezing point of food to -18 °C. Frozen foods last many months without spoiling however, some quality loss may occur.

Some microorganisms grow even at sub-freezing temperatures as long as water is available. Conversion of water to ice increases the concentration of dissolved solutes in unfrozen water and leads to low water activity. Freezing prevents the growth of microorganisms due to reduced water activity. The concerted effect of low temperatures, reduced water activity, and pre-treatment of blanching prior to freezing of products yield longer shelf life.

Different types of freezing systems are available for foods. No single freezing system can satisfy all freezing needs, because of the wide variety of food products and process characteristics. The selection criteria of a freezing system will depend on the type of the product, reliable and economic operation, easy cleanability, hygienic design and desired product quality.

Although all commercial freezing processes are operated at atmospheric conditions, there are potential applications of high-pressure assisted freezing and thawing of foods. The pressure–induced freezing point and melting point depression enables the sample to be super cooled to low temperature (e.g. -22 °C at 207.5 M pa) resulting in rapid and uniform nucleation and growth of ice crystals on release of pressure.

Freezing Systems based on Time taken for Freezing

Freezing systems based on time required to freeze foods can be classified into two types i.e. slow and quick freezing. Rate of freezing affects the quality of frozen food.

- Slow Freezing: Slow freezing occurs when food is directly placed in freezing rooms called sharp freezers. It is also known as sharp freezing. This method involves freezing by circulation of air by convection i.e. through a specially insulated tunnel, either naturally or with aid of fans. The relatively still air is a poor conductor of heat and that is the reason for long time required to freeze the food. The temperature ranges from -15 to -29 °C and freezing may take from 3 to 72 hours. The ice crystals formed are large and found in between cells i.e. extra-cellular spaces because of which the structure of food is disrupted. The structure of food is not maintained and thawed food cannot regain its original water content. Large ice crystals create quality problems like mushiness in vegetables etc.
- Quick Freezing: Vigorous circulation of cold air enables freezing to proceed at a moderately rapid rate. In this process, the temperature is kept between -32 °C to -40 °C and the food attains the stage of maximum ice crystal formation in 30 minutes or less. Small ice crystals are formed within the cells and therefore, it does not damage the structure of food. On thawing, the structure of original food is maintained.

Slow vs. Quick Freezing

The difference between sharp and quick freezing has been given in Table.

Table: Difference between sharp and quick freezing.

Sharp freezing	Quick freezing
• Rates of cooling of less than 1 °C/min. Ice crystals form in extra-cellular locations. • Large ice crystals formation. • Maximum dislocation of water. • Shrinkage (shrunk appearance of cells in frozen state). • Less than maximum attainable food quality.	• Produces both extra-cellular and intracellular (mostly) locations of ice crystals. • Small but numerous ice crystals. • Minimum dislocation of ice crystals. • Frozen appearance similar to the unfrozen state. • Food quality usually superior to that attained by slow freezing.

Freezing Systems based on Mode of Operation

They are constructed for batch, continuous and batch/continuous modes of operation.

- Batch Freezing: This type of freezing is mostly used for small operations. If a variety of products are to be frozen, a batch freezer may be selected over continuous as they are more versatile. Such freezing systems are also likely to be used for products with longer freezing times since with a batch freezer there is better utilization of floor space due to the multi-layer arrangement of loading. It is difficult to choose on an exact line of demarcation but generally freezing times longer than one hour would usually require a batch mode of operation.
- Continuous Freezing: This process of freezing is used in large-scale production lines. Continuous freezers are best used for freezing individual portions, such as small pieces of vegetables. The main advantage in using a continuous freezer for these smaller and/or thinner products is that since they freeze quickly they will also thaw quickly and the delays that occur with a batch-freezing operation may be overcome. Continuous freezing allows quick handling after freezing and a quick transfer to the cold store.
- Batch/Continuous Freezing: These kind of freezers are usually batch type freezers operated with trolleys which are loaded in sequence at fixed-time intervals rather than all at one time as in the truly batch freezer.

Freezing Systems based on Contact with the Product

Based on the contact between food and the freezing medium, the freezing systems can be categorized in two categories viz. direct and indirect contact freezing:

- Direct Contact Freezing: In case of direct contact freezers, the product to be frozen is completely surrounded by the freezing medium i.e. the refrigerant, which maximizes the efficiency of heat transfer.
- Indirect Contact Freezing: In this type of freezer, food materials to be frozen are separated from the refrigerant by a conducting material, usually a steel plate/ belt. The food product is indirectly exposed to the freezing medium.

Freezing Systems based on Heat Transfer Systems

Different types of freezing methods based on the heat transferring systems are:

Air

In general, air is used as the freezing medium in the freezing, either as still air or forced air. Packaged or unpackaged solid foods can be frozen in air at temperatures ranging from -18 to -40 °C. Different methods are available for freezing by contact with cooled gas.

Cabinet Freezing

This method involves circulation of cold air in a cabinet where the product is placed in a tray. The moisture pick-up from the product may deposit on the cooling coils as frost, which acts as an insulation.

Air Blast Freezing

This method employs the use of high velocity cold air to reduce the temperature of the food. Air velocities between 2.5 and 5 meter per second give the most economical freezing. Lower air velocities result in slow product freezing and higher velocities increase unit freezing costs considerable.

This method can be further divided into fluidized bed freezing, belt freezing, and tunnel freezing, depending on how the air interacts with the product.

Belt Freezing

The first mechanized air-blast freezers consisted of a wire mesh belt conveyor in a blast room for continuous product flow. Uniform product distribution over the entire belt is required to achieve uniform product contact and effective freezing. Controlled vertical airflow forces cold air up through the product layer, thereby creating good contact with the product particles and increasing the efficiency. The principal current design is the two–stage belt freezer. Temperatures used are usually -10 to 4 °C in the pre-cool section and -32 to -40 °C in the freezer section.

Tunnel Freezing

In this process, the products are placed in trays or racks in a long tunnel and cool air is circulated over the product.

Fluidized Bed Freezing

The fluidized bed freezer consists of a perforated bed through which refrigerated air is blown vertically upwards. The air velocity must be greater than the fluidization velocity. A schematic diagram of fluidized-bed freezer is shown in this freezing method is suitable for small particulate food bodies of fairly uniform size, e.g. peas, diced carrots, corns and berry fruits. The high degree of fluidization improves the heat transfer rate and results in good use of floor space.

The use of fluidization is advantageous compared with other methods of freezing since the product is individually quick frozen (IQF). The idea of individually quick frozen foods (IQF) started with the first technological developments aimed at quick freezing for the food particles with a tendency to stick together. Small vegetables, prawns, shrimp, french-fried potatoes, diced meat, and fruits are some of the products now frozen with this technology.

Plate Freezing

In this method, the product is sandwiched between metal plates and pressure is usually applied for good contact. Most familiar types of such freezers include vertical plate freezers and horizontal plate freezers based on loading of food in freezer. Plate freezers are only suitable for regular-shaped food materials such as blocks, packages and cartons. Batch freezing is the most common for plate freezers, however some are available for a continuous operation. When the product has been frozen, hot liquid is circulated to break the ice seal and defrost. Spacers are used between the plates during freezing to prevent crushing or bulging of the package. Temperatures are between -17 °C to -45 °C.

Immersion Freezing

In this method, the food is immersed in low-temperature brine to achieve fast temperature reduction through direct heat exchange. The refrigerating fluids usually used are salt solutions (sodium chloride), sugar solutions, glycol and glycerol solutions, and alcohol solutions. These should have low boiling points. Immersion freezing is advantageous as compared with air blast freezing in terms of reduced weight loss by evaporation.

Refrigerant must be non-toxic, pure and should not contribute to taste or flavour of food. Covering food with flexible membranes, which allow rapid heat transfer, may be done to prevent contact between food and liquid refrigerants. Immersion freezing with the use of liquids has now only a limited use. Nowadays, liquefied gases have replaced the salt/ sugar solutions and are used mainly for continuous freezing processes.

Cryogenic Freezing

Cryogenic freezing, is one of the fast, flexible and cheap method of freezing. In this, the liquefied gases are placed in direct contact with the foods. The three liquefied gases most commonly used as refrigerants are nitrogen, carbon dioxide and liquid freon freezant (CCl_2F_2). Unlike nitrogen and carbon dioxide, most of the refrigerant is recondensed and used again, resulting in reduction of cost. The product can be exposed to a cryogenic medium in three ways:

- The cryogenic liquid is directly sprayed on the product in a tunnel freezer.
- The cryogenic liquid is vaporized and blown over the food in a spiral freezer or batch freezer.
- The product is immersed in the cryogenic liquid in an immersion freezer.

Food is exposed to an atmosphere below -60 °C through direct contact with liquid nitrogen or liquid carbon dioxide or their vapour. This is a very fast method of freezing, thus adequate control is necessary for achieving quality products. It also provides flexibility by being compatible with various types of food products and having a low capital cost. The rapid formation of small ice crystals greatly reduces the damage caused by cell rupture, preserving colour, texture, flavour and nutritional value. The rapid freezing also reduces the evaporative weight loss from the products, provides high product throughout and has low floor space requirements. The advantages of liquid nitrogen are that it is colourless, odourless and is chemically inert and boils at -195.8 °C. It is usually used for high-value products due to the high capital cost for gas compression.

Dehydro-freezing

Dehydro-freezing is a process involving dehydration followed by quick freezing. First of all, food product is removed of its 50-60 per cent of the water by evaporation before freezing. The texture and flavour of food products frozen using such process are claimed to be superior to those resulting from freezing alone, and rehydration is more rapid than with dehydrated products.

Quality Loss due to Freezing Temperatures

Freezing is done to preserve food by reducing the product temperature, thereby slowing the quality deterioration processes. But still some changes take place on long storage of frozen foods like oxidation of fat, growth of microorganisms, enzymatic reactions and loss of surface moisture (dehydration) which needs attention.

Chemical and Physical Changes

Many changes, both chemical and physical may take place during freezing and subsequent frozen storage. Enzymes may catalyze chemical change detrimental to the quality of the frozen food. The action of lipase is more likely to be a problem than the activity of proteases during frozen storage.

The treatment of the food prior to storage influences the extent of enzyme activity subsequently. Enzymes, which cause the loss of colour, loss of nutrients and flavour changes in frozen food especially vegetables, must be inactivated to prevent such reactions from taking place. Since fruits are not blanched, enzymes in fruits to be frozen should be controlled by using chemical compounds such as ascorbic acid (vitamin C), which interfere with deteriorative chemical reactions. Other temporary measures include soaking the fruit in dilute vinegar solutions or coating the fruit with sugar and lemon juice to control enzyme-activated browning. But these are not as effective as ascorbic acid.

Oxidative rancidity may develop in frozen products through contact of the frozen product with air. To prevent this problem, proper packaging which does not permit air to pass into the product should be used. Physical changes include clumping of egg yolk.

Flavour Changes

Flavour may be modified during frozen storage as a result of enzymatic or oxidative changes in the food. Vegetables and food containing fats are most susceptible to the detrimental changes in flavour during frozen storage. The extent of flavour modification is determined by treatment prior to freezing and by storage condition during storage. Vegetables need to be blanched prior to being frozen to aid in minimizing flavour changes. Blanching helps to inactivate lipoxidase, the enzyme which causes development of rancidity in fats in vegetables. Damage to vegetables results in development of off flavours if these vegetables are held for a storage period prior to processing.

Textural Changes

Texture of frozen foods is modified by the freezing process. Whether the change is desirable or not will depend on the food itself. When the water freezes, it expands and the ice crystals cause the cell walls to rupture. As a result, the texture of the produce after thawing will be markedly much softer than it was when raw. Therefore, frozen foods to be eaten should be partially thawed. Textural changes due to freezing are less noticeable in high starch vegetables, such as peas and corn etc.

Colour Changes

Pigment changes can be noted in frozen beef. The iron in the myoglobin can undergo an oxidative reaction to form metmyoglobin, a reddish brown pigment with less visual appearance than the original myoglobin or oxymyoglobin. Some of the reddish fruits and vegetables become less attractive during frozen storage because the loss of some of their anthocyanin pigments into the medium surrounding them. In peaches, enzymatic action can produce colour changes in storage. The polyphenol oxidase enzyme in peaches catalyses a chemical change of the flavonoid pigments with molecular oxygen resulting in formation of quinones which in turn are converted to brown coloured polymers yielding a familiar brownish colour of frozen peaches.

Moisture Loss

Fast cooling and freezing can prevent moisture loss or dehydration by minimizing the water evaporation rate and time.

Freezer Burn

The grainy, brownish spot formed where the tissues become dry and tough in frozen storage is known as freezer burn. This imparts off flavours to the product. Freezer burn can be prevented by using thick moisture proof packaging.

Nutritional Value of Frozen Foods

Freezing, when properly done, may potentially preserve the greatest quantity of nutrients. Nutritional merits of frozen foods are influenced by the preparation required prior to freezing. Blanching is necessary to inactivate enzymes but it also destroys Vitamin C to the extent ranges of 40 per cent. B-vitamins are also destroyed. Water soluble vitamins are lost in the drip while thawing frozen meats. This loss ranges between 3-10 per cent.

Curing and Pickling

Curing is a preservation method usually done for meat and fish. It is carried out by the addition of combination of salt, nitrite and/or nitrate salt, sugar. Curing is an appropriate method to prevent the microbial growth, preserve the original flavors of the meat, increase the water retention and fat binding properties, and also to keep essential minerals and vitamins. The curing is further continued by other specific processes such as fermentation, drying and ripening of meat. The length of curing process depends on the type of meat cut, size of the meat pieces and composition. The whole process can last up to three to six months. For few specialty ham products, the process may go up to 24 months.

Science behind Curing

Sodium chloride (common salt) or sodium nitrate or combination of these salts can be used for curing of meat or fish products. The addition of salt does not destroy the bacteria, but it prevents or slows down their growth by creating an unfavorable conditions. Instead of inhibiting the growth of all organisms, it allows for a moderate amount for fermentation to occur. The fermentation ends up with a creation of distinct flavor to the meat. The curing salt makes the meat or fish tender by break down of protein in muscle fibers. Finally, salt also act as a dehydrating agent, which removes water from the meat by diffusion. Hence the final product will contain less water and more salt. Thus the salt creates an environment which is detrimental to the microorganisms.

Curing Ingredients

The most commonly used curing ingredients are salt (sodium chloride), sugar, nitrite and/or nitrate and phosphates. Each ingredient has it unique characteristics and also plays an exclusive role in preservation.

Sodium Chloride

Sodium chloride or common salt is the basic ingredient in all curing methods. Salt serves as a preservative agent and flavoring agent to the cured products. It has the capability to bind water and withdraw mositure from meat. When meat is exposed to brine or salt, the free moisture is attracted towards the sodium chloride ions and reduces the water activity. Reduced water activity creates an unfavorable condition for the growth of microorganisms. The water holding capacity of meat is increased with addition of 5 % of salt in meat. This 5% of salt inhibits the growth of anaerobic bacteria whereas 10% salt inhibits almost the growth of most bacteria except few salt-tolerant species. The addition of salt also improves the flavor of cured meat. Based on the consumer acceptability in cured hams, the optimized salt concentration is found to be 2.5%.

Nitrate/Nitrite

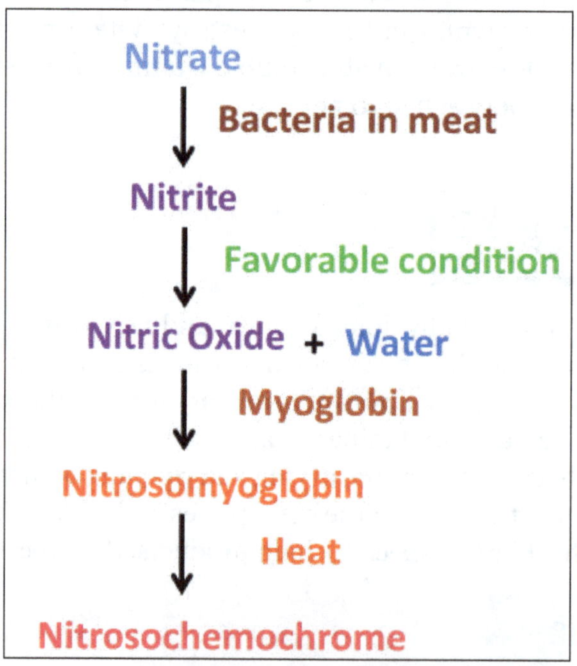

Nitrates and nitrites are added during curing in the form of potassium or sodium nitrates and nitrites. They are toxic when used in large amounts. Hence it is mandatory to use the exact amount of the above curing agent. The major functions of nitrate/nitrites is to stabilize color in cured meats, to impart characteristic flavor to meat, to retard rancidity in meat by stabilizing the fat and to inhibit many pathogenic organisms especially Clostridium botulinum. Sodium nitrate does not directly involve in curing, it get converted into nitrite which is responsible for the further reactions. The bacteria present in meat react with nitrate to form nitrite. Sodium nitrite has the ability to react with the red meat pigment to form the heat stable red curing color. A mixture of 99.5% salt and 0.5 % sodium nitrite is available to reduce the risk of overdosing. This homogenous mixture is called as curing nitrite salt. Nitrogen oxide is formed when nitrite is exposed to acidic environment. This nitrogen oxide combines with myoglobin to form nitrosomyoglobin which is a heat stable, bright red compound. A 2% of nitrite in curing salt generates approximately 150ppm (150 mg/kg) nitrite in meat products. On further reaction with myoglobin, the cured meat product will contain 50-100ppm of residual nitrite. This residual nitrite in cured meat should not be exceeding 125ppm.

Sugar

Sugar or corn syrup can also be added during the curing to reduce the harshness of salt and excess moisture removal. This process is also called as sweet pickling. The sugars most frequently used are sucrose, cane sugar, dextrose, and invert sugar. Addition of sugar also imparts flavor to the cured meat by serving as a food by flavor producing bacteria. The presence of sugar in the cured meat imparts browning during smoking and a characteristic flavor during cooking. Sugar prevents the oxidation of meat pigments by blocking the formation of unwanted derivatives during curing process. It also improves the color of meat pigment. It provides food for the desirable microorganisms which produce acid and favor for the production of necessary meat pigments.

Phosphates

The addition of phosphates during wet curing of meat tends to double the water binding capacity of meat products. It results in the increase in pH which increases the capacity of the muscles to bind water. The increased pH causes unfolding of the muscle proteins, thereby making more sites on the proteins available to bind water. Sodium tripolyphosphate, sodium hexametaphosphate and sodium pyrophosphate are approved by USDA for use in meat curing.

Curing Accelerators

The duration of curing can be shortened by the addition of curing accelerators. The usage of reducing compounds (reductants) and acidulants are important for the development of color and other meat properties. Ascorbic acid, sodium ascorbate, erythorbic acid and its derivatives are the reductants widely used for meat curing. The major role of curing accelerators is to speed up the chemical conversion of nitrite to nitrosomyoglobin to obtain the stable meat pigment. These accelerators are added along with the curing salt. Sometimes, the curing agents are coupled with acidulants such as fumaric acid and glucano-delta-lactone for rapid conversion of nitrate to nitrite.

The reductants such as ascorbic acid or vitamin C and isoascorbic or erythorbic acid possesses antioxidant properties. Sodium ascorbate and sodium erythorbate are sodium salts of the acid are commonly used in meat curing. They are effective in reducing metmyoglobin to myoglobin and accelerate the production of nitric oxide from nitrite. They are highly effective in maintaining the fresh meat color. The ascorbic acid or erythorbic acid is added in cured meat or poultry at a maximum concentration of 469 ppm.

The acidulants provide an acidic environment which encourages faster conversion of nitrite to nitric oxide. Reduced pH will accelerate the curing process but also reduces the water binding capacity of the meat which in turn reduces the yields. A maximum concentration of only 0.5% is permitted to add in the meat.

Methods of Curing

There are three methods of curing. They are dry curing, wet curing and injection curing.

- Dry Curing: Dry curing is also referred as dry salting or corning. The term 'corning' comes from the word 'corn' which means grains of salt. It is the process of applying dry salt along with sodium nitrate(99.5% common salt or sodium chloride and 0.5% sodium nitrite). The whole meat pieces or meat cuts are rubbed with the salt and stacked on each other with layers of curing salt. The salted meat cuts are stored at low temperature (0 to +4 °C). The product is completely covered to protect from the air. During this process, the salt diffuses through the meat tissues and the liquid from meat oozes out. The extracted liquid from the salted meat is collected separately from the bottom of the stacks. Dry curing mix can be sprinkled over the piled products. For uniform curing, rearranging of the lower pieces to the upper stack and vice versa should be done. Dry curing is especially carried out sausages, hams and bacon. These dry cured products will be further smoked and air dried, which does not need additional cooking.
- Wet Curing: The wet curing method otherwise called pickling or immersion curing. This process involves the immersion of meat in the brine solution. The brining solution contains water, salt and optional spices or herbs. The meat pieces kept in the tank filled with brine and immersed completely in the solution. The temperature of the curing tank will be maintained around 8 to 10 °C. the duration of curing may range from several days to 6 weeks depending on the size of the meat. For larger meat cuts, due to their lengthy duration of curing, there might be danger of spoilage on the products. For smaller meat cut, it takes 3 – 14 days, hence it is recommended for smaller cuts of meat.
- Injection Curing: Three forms of injection curing are possible. They are stitch or spray pumping, multi needle machine injection and artery pumping. The brine is prepared by mixing salt with water, curing agent such as nitrate, nitrite and other additives like spices, herbs, sugar, soy protein etc. The addition of alkaline phosphate to the brine is also recommended which increases the yield of final product.

Stitch or Spray Pumping

In this method, the brine solution is injected directly into the meat using single orifice needle. Through this method, the solution is rapidly distributed in the muscle channels. The stitch method is used for meat cuts with uneven surfaces. Around 3 - 5 stiches per cut of meat is made to bring the weight of the brine to the added level. Spray pumping is a modified method where the needles with equal spaced holes along the length of the needle for uniform distribution of brine into the muscles. The injections are made at several places as close together as possible. The stitched meats are kept under refrigeration condition for 5-7 days to allow uniform distribution.

- Multi Needle Injection Curing: The curing process can be automated multi needle injector. The automatic injector consists of storage tanks, brine mixers, feed hoppers, multi needle injectors and continuous massagers. After injecting the brine solution, the meat products are rotated inside tumbling equipment for uniform spreading of the solution. The quantity of brine to be injected into the fresh meat can be preset by fixing the pressure and speed. The needles and hoses are thoroughly cleaned with warm water and disinfected before starting of the process. Any microorganisms remaining in the system would be injected deep into the meat pieces during the operation, hence absolute cleanliness is essential.

- Artery Pumping: In this method, the brine solution is pumped into the meat cuts through arterial system. This method is suitable for meats with intact vascular system like hams. The brine solution is pumped in to the meat cuts using needle connected to pump at a pressure of 275- 345 kPa. Usually arterial pumping adds 8-10% of brine by weight of the final product. For equilibration of the cure and for fixation of meat color, the cured product should be maintained at refrigerated temperature for about 24 hours. Careful handling is required while pumping the cure liquid to avoid bursting of arteries.
- Thermal or Hot Cures: This method utilize hot brine or hot dry cure mixtures. The hot brine solution is injected into the meat cuts either using stich pumping or artery pumping. Hot curing solution accelerates the curing process. Another case, where perforations are made in the meat surface using needles and the hot dry cure mixture is applied and left for 3-5 days at 9-10 °C. The hot curing process is rapid, allows greater penetration of smoke during smoking through the surface perforations, increases the yield than dry cures and also improves the flavor of meat. The only limitation is that this process can be applied to thin and uniformly shaped products.

Pickling

Pickling is also a method of food preservation mostly occurs in an acidic environment such as brine or vinegar. Pickling is more common with vegetables. Proper pickling extends the shelf of the product and also inhibits the growth of microorganisms. During pickling the product is usually immersed in brine or vinegar. In regions of the world such as in East Asia vegetable oil and vinegar is used as the pickling medium. The pickles add palatability to the meal and tend to stimulate the gastric juice and thus help in digestion.

Science behind Pickling

If the raw material has sufficient moisture content, then the brine solution is produced just by the addition of dry salt. During pickling, the brine solution penetrates into the tissues of vegetables and soluble material present in them diffuses into the brine by osmosis. Pickles are the results of lactic acid fermentation of vegetables. The lactic acid bacteria present in fruits or vegetables suppress the growth of spoilage microorganisms during pickling by creating an unfavorable acidic environment. Most of the spoilage microorganisms cannot withstand brine containing 15% salt. Lactic acid bacteria being salt-tolerant can thrive in brine of 8-10 per cent strength though fermentation takes place fairly well even in 5 per cent brine. For vigorous lactic acid fermentation, the recommended concentration is 10% salt solution. The production of lactic acid preserves the vegetable and gives fermented pickles their characteristic tartness and flavor.

Pickling Methods

Pickles can be prepared using dry salt, vinegar, vegetable oil or the combination of the above with herbs and spices.

Dry Salted Pickles

Dry salting can be applied to many fruits and vegetables like lime, lemon, mango, cucumber. The vegetable to be pickled is treated with dry salt which extract juice from the product and produces brine. Salt content of 15 % or above prevents microbial spoilage and controls fermentation. It also improves the flavor and taste of the final product. Usually spices are added to the pickles. Fermented pickles are produced with salt.

For the preparation of dry salted lime pickle, one layer of cut limes are placed in a vessel and covered completely with salt. Similarly another layer of cut limes and salt are placed. This layering process is continued till three fourth of the vessel is filled. A cloth is covered the top layer and heavy weight is placed over the cloth to compress the products. The pressure given at the top assists in the extraction of juice from the product. The extracted juice mixed with the salt and thus produces brine. Fermentation starts as soon as the brine is formed and bubbles of carbon dioxide begin to appear. A temperature of 21 °C is ideal for lactic acid fermentation; hence the vessel is placed in warm sunny place to allow fermentation for a week. Absence of bubble production indicates that the fermentation is ceased. To the fermented lime, seasonings like roasted mustard, cumin seeds, fenugreek seeds, chili powder and turmeric powder can be added as per the taste.

Preservation with Vinegar

Vinegar serves as a preservative and flavoring agent to pickles. The cut vegetables are immersed in strong vinegar of about 10 % strength. This prevents the dilution of vinegar by the water extracted from the vegetable and also expels the gas present in the intercellular spaces. Salt, sugar and spices are added to flavor these pickles. The vinegar concentration in the final product, in terms of acetic acid should not be less than 2 %.

Vegetables pickled with vinegar gives a crunchy texture to the product. Cauliflower, onion, garlic, cucumber, celery, carrots and zucchini can be pickled using vinegar. The vegetables are blanched for about 1-2 minutes in brine solution before transferring to the jar. Sometimes the vegetables are boiled along with vinegar for about 15 minutes. The jar is cooled to room temperature and the product should be stored in a cool dry place. Along with the vinegar, bay leaves, coriander, mustard seeds, herbs such as rosemary, thyme, basil can also be added.

Preservation with Oil and Spices

Oil is a popular medium used for preserving pickles in our country. It simply seals off the air from the product. The spices are added along with oil. The oil also helps the seasonings to adhere better to the product.

For the preparation of mango pickle with oil, the cut mangoes are dried just to remove the surface moisture. The pickle mix is prepared by adding salt, chilli powder and mustard seeds together. Usually vegetable oil or mustard oil is used for pickling. The oil is heated and the pickle mix is added to the oil once it is cooled. The mix is blended well with the oil without any lumps. To this mixture the cut mangoes are added along with fenugreek seeds, asafetida and turmeric powder and mixed well. The blended product is transferred to sterile airtight jar. The pickle should be stirred every alternate days and it will be ready within a weeks' time for consumption.

Smoking

Smoking is one of the ancient preservation method used to preserve cured meat and fish by using the products of smoke. This process also increases their palatability by adding flavour and color to the smoked product. The drying action of the smoke tends to preserve the meat through the chemicals present in wood smoke are natural preservatives as well. Though the cheeses, vegetables, beer ingredients were smoked, the most commonly smoked products are meat and fish. Smoking is considered as a complementary process to curing. Smoking was used to preserve meats by absorption of aromatics from the smoke. Certain component of smoke acts as a powerful preservative of meat. The smoked meats are less susceptible to rancidity and mold growth. Smoked meats and poultry are widely accepted for their peculiar smoky flavor than for their preservation effect.

Types of Smoking

The following are the few types of smoking:

Cold Smoking

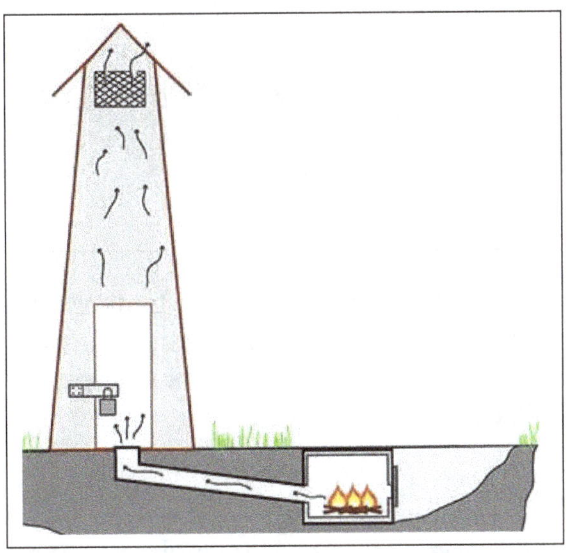

Cold smoking is an old process in which, the smoke is passed through pipes to the meat which is kept in a separate chamber. The smoke gets cooled while passing through the pipe before it reaches the meat. During this type of smoking, the meat is kept very close to room temperature and hence it takes a longer duration.

As per Food and Drug Administration (FDA), the product is not cooked during cold smoking and also the temperature of air around the food product is kept well below 30°C. This represents that the food is in temperature danger zone where microbial growth may occur. Hence the productswhich are salted, cured and fermented are cold smoked. Salting or curing of meat ensuresthat bacteria will not develop while it is smoked and stored. The more intense smoke flavor in meat indicates that they are left longer in the smoker. As the cold smoked meats are not cooked, they tend to have smooth texture which often reminds the texture of uncooked food product. This type of smoking does not preserve the product and should be refrigerated till consumption. It is highly recommended to cook the cold smoked meat products before consumption. Cold smoking is effectively carried out in the winter season because during hot season, the temperature inside the smoker may rise high.

Hot Smoking

Hot smoking is a process in which the meat product is enclosed in a smoker and directly exposed to smoke from wood. This process involves heating and high temperature of the product.

Hot smoked products does not need further cooking. The temperature of the hot smoked product ranged from 52 to 80°C. For hot smoking, modern electric kilns are used in which the meat is cooked and smoked at the same time over a burning fire or electric elements of a kiln. It takes only short period of time to smoke the product.

Smoke Roasting

The combination of smoking and roasting is known as smoke roasting. This method is also referred as barbecuing or pit baking. It may be done in a closed wood-fired masonry oven or barbecue pit where the temperature of the product is reached above 121 °C.

This type of roasting can be performed in any conventional oven by keeping a pan filled with hardwood chips on the floor of oven so that the chips burn and produce a smoke. This type of smoking should be carried out in a well-ventilated area to prevent carbon monoxide poisoning.

Pan Smoking

It is simple smoking method when smoker or smokehouse is not available. It is an inexpensive method which imparts the smoky flavor to foods in a short duration. This process requires two disposable aluminum pans, a rack and sawdust. The drawback of is that the smoke cannot be controlled and sometimes the products tend to get very intense smoke flavor which results in bitter taste.

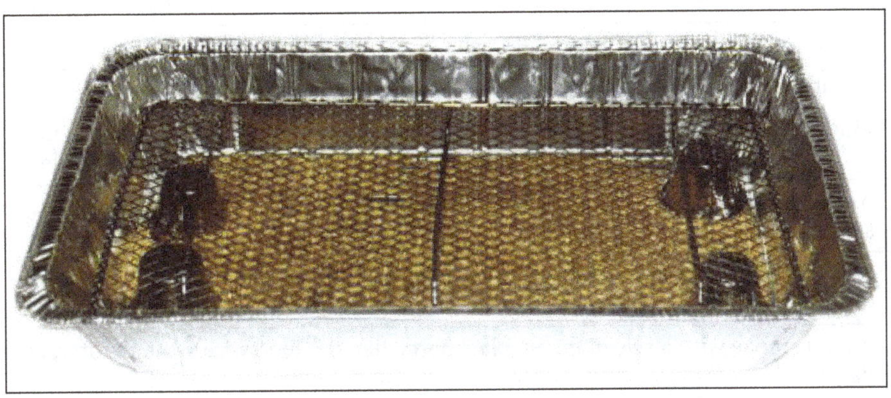

Liquid Smoke

Liquid smoke is obtained by controlled burning of wooden chips in the combustion chamber which produces the smoke. It is then passed through a tube from a combustion chamber to a condenser. In the condenser, the smoke cools and condensed to form a liquid. The liquid smoke is collected, filtered to remove impurities and then bottled. This method of distillation produces wide of smoke flavors. Currently liquid smoke has been used in many commercial operations to add smoke flavor to the food products. Liquid smoke has advantages over traditional smoking is that it can be controlled and immediate smoke flavor is attained.

Selection of Wood for Smoking

The heavier hardwoods like oak, hickory are used for smoking "heavier" meats like beef and pork whereas, lighter hardwoods like alder, maple, fruit and nut-bearing woods are used for smoking "lighter" meats like poultry and fish. The softwoods like pines, cedar are not preferred for smoking meat, as they burn fast and produce significant quantities of resin on burning. They also impart undesirable taste to the food. Burning of fresh cut hardwood produces lot of steam and off flavour during combustion due to high moisture of about 50% by weight. Such woods will take up to 45% more energy than charcoal or gas. Hence for this reason, dried hardwoods are preferred for smoking of foods. Dried hardwood has mositure around 5% and the remaining 95% constitute the following components: cellulose (40%), hemicellulose (40%), lignin (19%), and minerals (1%). Also the moisture content will vary depending on the wood variety, age, soil, and climate.

Table: Wood smoke flavor chart.

Wood Type	Smoke Characteristics	Goes good with products
Almond	Nutty and sweet smoke flavor, light ash.	Good with all meats.
Apple	Slightly sweet but denser, fruity smoke flavor.	Beef, poultry, game birds, pork.
Maple	Mildly smoky, sweet flavor.	Goes well with poultry, ham, cheese, small game birds and vegetables.
Oak	A mild smoke with no aftertaste. Also gives a good smoked color.	Good with red meat, pork, fish and big game.
Pear	Smoke flavor much like apple. Slightly sweet, woodsy flavor.	Good on Poultry, game birds and pork.
Mulberry	A mild smoke with a sweet, tangy, blackberry-like flavor.	Good with Beef, poultry, game birds, pork.
Lime	Medium smoke flavor with a hint of fruitiness.	Excellent with beef, pork and poultry.

Wood Smoke

Wood smoke is an aerosol produced by pyrolysis of wood at elevated temperatures and reduced oxygen. It is a combination of tiny particles mixed with water vapor and a complex cocktail of gases. The sole fuel source in smoking of foods is the dried hardwoods. Hardwoods consistedof cellulose, hemicellulose and lignin, in which cellulose and hemicellulose are the importantstructural material of the wood cells. On burning of wood, cellulose and hemicellulose caramelize and produce carbonyls. Lignin, a complex arrangement of interlocked phenolic molecules, is the cell-bonding substance present in wood cells. When burnt, lignin also produces a number of characteristicaroma compounds. The minoramount of proteins in wood contributes to the roasted flavors. As the phenolic compounds in smoke are unstable, they dissipate after a few weeks or months from the product.

Composition of Wood Smoke

Wood smoke is composed of a gas, particles of liquid droplets and solids. There are over 400 compounds were identified in different sources of wood smoke. Until now, 131 carbonyls, 75 phenols, 46 furans, 40 acids, 16 lactones22 esters and alcohols wererecognized. Thus, the composition of wood smoke depends on the type of wood, temperature and humidity of smoking, oxygen availability and moisture content of the wood. The carbonyls produced during burning are responsible for the color components and sweet, fruity aromas. The burning of lignin in wood produces pungent smokyflavor components such as guaiacol, phenol and syringol. They also providesweeter scents such as vanilla-scented vanillin and clove-like isoeugenol. Guaiacol is the main phenolic compound contributes to the smoky taste, whereas syringol is the primary contributor to smoky aroma. Table represents the predominant components of wood smoke.

Table: Predominant compounds in wood smoke

Class of compound	Examples
Acids	Acetic acid
Alcohols	Methanol, ethanol, isopropanol, furfuryl alcohol
Carbonyls (aldehydes)	Acetaldehyde, isovaleraldehyde, crotonaldehyde, diacetyl
Carbonyls (ketones)	Acetone, methyl vinyl ketone, acetol, 2-cyclopentnone, furfural, 5-ethyl furfural, cyclotene
Esters	Methyl formate, methyl acetate
Furans	Furan, 2-methyl furan
Phenols	Phenol, 1,2-dimethoxybenzene, guiacol, 4-methyl guiacol, 4-ethyl guiacol, eugenol, syringol, 4-methyl syringol, 4-ethyl syringol, 4-propyl syringol, vanillin

Functional Properties of Smoke

The functional properties of smoke components are flavor, color and antimicrobial activity. The peculiar smoky flavor is due to the presence of phenolic components. The blend of carbonyls and acids contributes to the flavor of smoke. Some of the phenolic components in smoke are similar to those present in spices. An example is eugenol in cinnamon, pepper, nutmeg, marjoram and cloves. The flavor of smoke components also depends upon the concentration. At high concentration, it provides sensory responses of burnt and pungent taste whereas at desired concentration, it produces sweet and smoky taste.

Wood smoke also acts as a colorant. A significant color is formed in the product when smoke and food components react chemically at the elevated temperature. The presence of acids and phenols are responsible for the coloring ability of wood smoke, while color formed during the heating of meat wasmainly due to the acids and carbonyl compounds. When the product is heated, the carbonyl compounds react with the proteins in a Maillard reaction to produce the brown color. The wood smoke also enables the formation of crust on meat. Figure shows the meat smoked by charcoal and gas, indicates the dark colored crust formation in charcoal smoked than gas smoked meat.

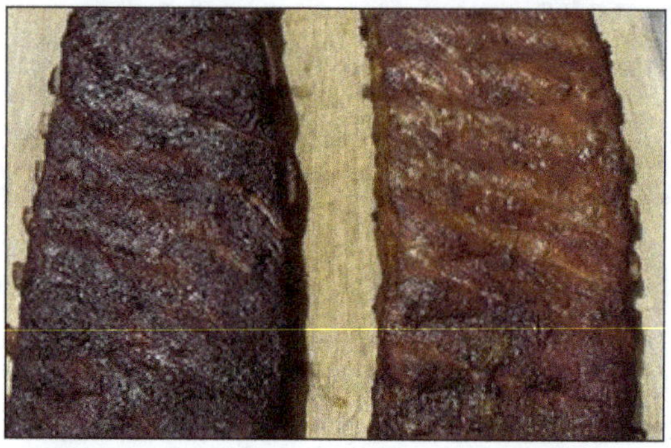

Charcoal smoked and gas smoked meat.

Among the functional components of smoke, phenols, acids and carbonylsare found to have antimicrobial activity. Carbonyls and acids have a wide spectrum of antibacterial activity even at low levels of phenols. Studies on antimicrobial properties of 0.5% smoke preparations was found to be bactericidal to pathogens such as Escherichia coli, Staphylococcus aureus and Listeria monocytogenes. Thus smoking acts as a preservative agent.

Types of Smokers

Smokers are available in different size and shape with wide variety of heating source. The different types of commercially available smokers are discussed below:

Vertical Smoker or Charcoal Smoker

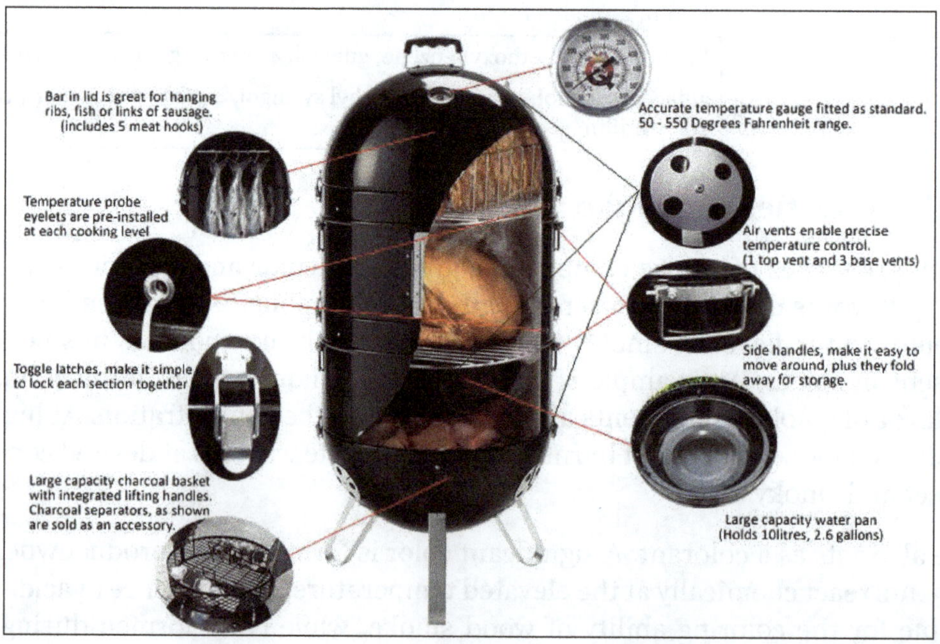

Vertical smoker is the mostly commonly used type due to its compact size, cost and ease of operation. It consists of 2 bowls, one is to place the charcoal and smoker wood and other bowl holds water or other liquid to keep the product moisturized throughout the smoking process. The front door of the smoker is to feed in new charcoal and wood. The product to be smoked is stored in the rack at the top. Hooks provided at the top can be used to hang the meat. The heat will rise straight up from the bottom bowl to the top.

Offset Smoker

The offset smoker consists of cylindrical cooking chamber attached with a firebox through a connecting pipe. Fire is lit in the firebox and airflow is tightly controlled. Apart from the vent in the cooking chamber, sometimes, the firebox is also provided with vent to regulate the temperature. The heat and smoke from the fire is moved through the connecting pipe into the cooking chamber and it flavors the meat. The excess smoke is vent through the exhaust pipe of the cooking chamber. This can be used as normal grill and also to smoke meat products.

Propane Smoker

A propane smoker is used for smoking of meat in a controlled environment. In a propane smoker, the heat is generated by a gas burner directly under a steel or iron box containing the wood or charcoal that provides the smoke. Few vent holeson the top of the box are provided on the steel box. It smokes instead of burning, due to less oxygen. Desired combination of wood and charcoal can be used and also this method consumes less wood.

Pellet Smoker

The pellet smoker or pellet grill consists of a pellet bin filled with the wood pellets. Wood pellets are natural hard wood sawdust compacted and extruded into small 1/4-inch round pieces. An auger is turned by a variable low speed motor, which draws the pellets from the hopper into an internal fire chamber where the pellets burn and products heat and smoke. Heat and smoke diffuse into the product by a continuously running blower fan. The auger speed can be adjusted manually.

Advantages of Smoking

The following are the advantages of smoking of foods:

- It adds flavor and taste to foods.
- It imparts pleasing color to the food product.
- It prevents fat from rancidity.
- It extends shelf life of the product.
- It kills certain bacteria and slows down the growth of other microbes.

Limitations of Smoking

Wood smoke possesses antimicrobial and antioxidant activity, but smoke alone is not sufficient for preserving food. It needs to be combined with other preservation methods like salting or curing or drying. Smoking protects the most vulnerable exterior surface of food by giving an extra protection whereas other preservation methods protect the interior of foods from microbial spoilage. There are chances for the food product to get dried due to long duration of smoking. The added problem faced during smoking of foods is that the smoke components stick to the external surfaces of the food and does not penetrate deep into the product. In modern cuisines, smoking is chiefly carried out for its characteristic flavor. Though artificial smoke flavoring is available in form of liquid smoke, it mimics the flavor of smoking but not its preservative qualities.

Smoked Food Products

The following are the few products that are commonly smoked to enhance the taste and flavour or foods.

- Smoked Meats: Smoked turkey, smoked chicken, beef and pork are flavorful options for preparing tastier meats. Smoking imparts a significant flavor and increases the shelf life of meats. To get delicate flavour of poultry meat, they are processed with light smoke. Addition of spices, seasonings and condiments also improves the flavors. Smoked meat is a method of preparing red meat. It is done to preserve the protein-rich foods, which would otherwise spoil quickly. Dehydration, antibacterial properties of phenols and other chemicals in the absorbed smoke preserves the product for a longer period. Smoked and cured sausages do not require refrigeration and further cooking before eating. The flavor of smoked foods makes them a delicacy in many cuisines.
- Smoked Fish: Fish has been cured earlier by smoking. Originally smoking of fish was performed to preserve the fish. In modern times, fish is smoked for its distinctive taste and flavour imparted by the smoke. The common fish varieties that are smoked are Salmon, mackerel, herring, whitefish and trout are the commonly smoked fish varieties. Hardwoods like oak, alder, hickory, birch, maple and fruit woods are best for smoking fish.
- Smoked Cheese: Cheese which is specially treated by smoke-curing called smoked cheese. As a result of smoke-curing process the original color of cheese changes to yellowish brown color. Any types of cheese such as Mozzarella, Cheddar, Muenster and Swiss can be smoked. The type of wood alters the flavor of the cheese. The most frequently used woods are fruit woods such as apple, cherry, pear and peach wood. Nut woods like almond, pecan can also be used to smoke cheese. Cold smoking of cheese is preferred because it is must, not to cook or melt the cheese. Smoking of cheese causes the milk fat to rise to the surface and form a preservative skin on the surface of the cheese. Long-term smoking imparts a darker color to the exterior surface of cheese.
- Smoked Eggs: Smoked eggs can be prepared either from hard boiled eggs or by smoking uncooked eggs in their shells. Pecan, cherry, oak woods are preferred for smoking of eggs. Cold smoking of eggs yield a soft product with excellent smokey flavor and also maintains the integrity of the egg, whereas, hot smoking over cooks the eggs and yield rubbery and dry product.
- Smoked Spices: A wide range of smoked spices and herbs are available in market. For example, the chilli peppers have been dried and smoked over an oak fire, then ground into a fine smoked paprika powder. These smoked chilli products add a powerful smokey aroma to the dish and yield a dark red colour.
- Smoked Vegetables: Smoking of vegetables such as onion, tomato, cabbage, corn is also common. During smoking of vegetables they are sprinkled with salt or spices along with oil. They are further processed into soup or can be added in other dishes. The smokey flavor adds an extra taste to the dish.

Chapter 4

Techniques of Food Safety

There are numerous techniques which are employed to ensure that the food is safe for consumption. A few such techniques are food irradiation, food extrusion and vacuum packing. The topics elaborated in this chapter will help in gaining a better perspective about these techniques related to food safety.

Food Irradiation

The term "food irradiation" may be applied to any process that exposes food either to electromagnetic radiation or to high-energy particles. Electromagnetic energy can be generated by radioactive isotopes, as in the case of gamma ray irradiation, or by bombardment of thin metal films with high-energy electron beams to produce radiation, as in the case of X-ray irradiation. Alternatively, a high-energy electron beam ("e-beam") can be directed at the food itself. In all of these cases, the radiation is absorbed by the food, and more particularly, by the microbial organisms in the food. This absorption disrupts the complex organic molecules of the microbes, either preventing the microbes from reproducing or killing them outright. The effectiveness of the treatment varies based on the type of radiation used, the intensity of the radiation, and the microbe in question. The relative advantages and disadvantages of the three forms of food irradiation (gamma ray, X-ray and e-beam) used today are discussed below:

Gamma Ray Irradiation

The simplest form of irradiation, at least in concept, is gamma ray irradiation. In this form of irradiation, the source of radiation is a radioactive element that emits photons in the gamma ray range of the electromagnetic spectrum. Gamma ray photons have a higher frequency (and there-fore energy) than either ultraviolet or X-ray photons. Gamma rays can penetrate a target food (or medical product) to a depth of several feet and reach microbial contaminants anywhere within that range.

While simple in concept, gamma ray irradiation can be difficult in practice. The first difficulty is selecting a radioactive source element. In addition to radiating gamma rays, many radioactive elements also produce alpha rays (helium nuclei), beta rays (high- energy electrons or positrons) and/or high-energy neutrons. Alternatively, they might decay into another radioactive substance that generates these other forms of radiation. The other forms of radiation are undesirable because they have the potential to make the target food (or medical product) radioactive. To date, the only radioactive isotopes approved as having the proper radiation profile are Cobalt 60 and Cesium 137, with only Cobalt 60 being actually used for food irradiation at the present time. These radioactive isotopes are produced by exposure of the ordinary element to a nuclear reactor core, and their availability may be conditioned on the continued availability of nuclear power.

Even after a source is selected, there are logistical complications in gamma ray irradiation. Radioactive elements do not have an "off" switch, nor do they come equipped with directional or intensity controls. Gamma rays can be contained by immersion of the source in a sufficient quantity of water, but the source must be removed from the pool in order to irradiate the target food. In order to prevent inadvertent gamma ray exposure, the source must be insulated from the outside world by several feet of concrete.

E-beam Irradiation

E-beam irradiation, though it uses the same term as gamma ray irradiation, is a completely different kind of treatment. High-energy electron beams are produced in an electron gun, a larger version of the cathode ray gun found in devices such as televisions and monitors. The electrons can be directed by a magnetic field to a target food. The term "irradiation" is really a misnomer, since the food not exposed to electromagnetic radiation or beta rays (electrons produced by a radioactive source). Nevertheless, the process has a similar effect to that of gamma ray irradiation. E-beam irradiation requires shielding as well, but nothing like the concrete bunkers used in gamma ray irradiation. The disadvantage of the e-beam is its short penetration depth (about an inch), preventing its application to many foods and limiting the amount of food that can be processed in bulk.

X-ray Irradiation

X-ray irradiation is a relatively new technique that combines many of the advantages of the other two methods. Like gamma ray irradiation, X-ray irradiation consists of exposing food to high-energy photons with a long penetration depth. In this case, however, bombarding a metal film with a high energy electron beam produces the photons, allowing the radiation to be turned on and off. The device is a more powerful version of the X-ray machines used in medical offices. The device still requires heavy shielding, although the amount of shielding required is less than that for gamma ray irradiation. No radioactive substances or by-products are used in, or result from, the process.

Regardless of form, food irradiation is fundamentally about how much energy is absorbed by the target food. It is helpful to have a measurement for what dosage of radiation will be required independent of the amount of food to be irradiated. For this reason, radiation doses are measured in kiloGray (kGy). A dosage of one kGy indicates that the target sample receives 1000 Joules (metric units of energy, abbreviated J) per kilogram of sample mass. When measuring the effect of radiation on the microbe population of food, it is useful to have a measurement that does not depend on the number of microbial organisms in a particular sample of food. For this reason, the effect of radiation on microbes is measured by a dosage called the D value. The D value is the dosage of radiation required to reduce the microbe population of a sample by 90%.If a particular organism has a D value of 0.5 kGy in a particular kind of food, then exposing a 1 kg sample of that food containing the organism to 500 J of radiation will kill 90% of the population of that organism. An additional amount of dosage equal to the D factor will reduce the remaining microbe population by 90%. Thus, exposing the sample in the example above to 1000 J of radiation would reduce the microbe population by 99%; 1500 J would remove 99.9% of the microbe population; etc. Varying the power of the source or the duration of exposure controls the amount of radiation the target receives. The energy of electron guns used for e-beams and X-rays is typically measured in electron volts (eV), units of energy convertible to J.

The D value will depend primarily on the type of food irradiated and the type of organism to be eradicated by irradiation. Generally, the more complex the organism, the more sensitive the organism will be to radiation, since the operation of complex microbes is easier to disrupt. Viruses, the simplest form of life, are most difficult to destroy. Many bacteria collapse into a dormant state known as a spore (as contrasted with the vegetative state) when conditions are unfavorable to growth (e.g., when the oxygen or temperature levels are too low). The D value for spores is higher than the corresponding D value for the vegetative state.

Other factors that affect the D value are the strain of organism involved, the state of the food (frozen or unfrozen), ambient oxygen and temperature. The difference between frozen and unfrozen food is particularly important, since one of the most effective ways of controlling microbial pathogens in food is to keep the food below a temperature at which the pathogen can grow. More radiation is required to kill microbes in frozen food, and the slight heating that results from incidental absorption of radiation by the food has the danger of raising the food to a temperature that would allow pathogenic organisms to grow. Consequently, temperature effects must be carefully monitored in most foods.

Effect of Irradiation on Foodborne Microbial Pathogens

One of the biggest problems with the acceptance of food irradiation in society is that few people are fully aware of how serious the problem of foodborne illness can be. In a way, this reflects well on our existing food safety measures, since it presumably indicates a great deal of consumer confidence in the quality of the food they purchase. Even given the relative safety of the food supply, the harm created by food-related illness is nothing short of staggering. There are five common food borne pathogens which can lead to loss in productivity. They are Salmonella, Escherichia coli, Listeria monocytogenes, Campylobacter jejuni and Vibrio. FDA and USDA have approved doses of 3.0 kGy for poultry, 4.5 kGy for other unfrozen meats, and 7.0 kGy for other frozen meats. Comparing those doses to the D value reveals the percentage of the microbe that will be killed by irradiation at the allowed dose.

1. Salmonella: Depending on strain of bacteria and other factors, the D value of Salmonella ranges from 0.4 to 0.8 kGy. At the 3.0 kGy dose approved for poultry, irradiation would kill over 99.9% of the most radiation-resistant strains of Salmonella.

2. E. coli: E. coli is even more radiation sensitive that Salmonella ; it has a D value ranging from 0.2 to 0.4 kGy. Exposure of beef to a 4.5 kGy dose would reduce the amount of E. coli in the sample by a factor of 100 billion. Considering that E. coli is not ordinarily present in beef, food irradiation could effectively eradicate the problem of E. coli in beef products.

3. Listeria: Listeria in beef, pork and lamb has a D value ranging from 0.40 to 0.48 kGy. Using the 4.5 kGy approved dose would reduce the Listeria population by a factor of one billion. Unfortunately, irradiation has not yet been approved for processed meat products such as hot dogs, a common source of Listeria.

4. Campylobacter: Campylobacter is one of the more radiation-sensitive bacteria, with the C. jejuni species having a D value ranging from 0.18 to 0.24 kGy. The approved poultry dose of 3.0 kGy would leave only one trillionth of the original Campylobacter population.

Considering that Camplylobacter 's effects are often chronic and difficult to trace back to food, food irradiation could prevent a great deal of food-related illness that might be undetectable and effectively unavoidable otherwise.

5. Radiation-resistant microbe: Most of the discussion thus far has been confined to the most common non-viral pathogens. For the sake of completeness, a few more organisms should be mentioned. C. botulinum , for example, is one of the more deadly pathogens (although not quite so virulent as Vibrio vulnificus); nearly 8% of hospitalized cases of botulism result in the patient's death. C. botulinum also happens to be quite radiation-resistant in its spore phase, having a D value of between 2 and 4 kGy. Irradiation, at least at conventional doses, would have a limited effect on C. botulinum.

Viruses also play a significant role in food-related illness, and they are much more resistant to radiation that other microbial pathogens. Since people exposed to a virus can often developimmunity to the virus, much of the incidence of food-related viral illness is found in children. Hepatitis A is the most notorious of the viral food-related illnesses. About 5% of all cases of hepatitis A are traced back to food, but since 50% of all hospitalized cases have an indefinite source, the actual number of food-related cases may be significantly higher. Again, irradiation at conventional doses is unlikely to have a significant impact on hepatitis A, and sufficient irradiation could have undesirable effects on the characteristics of infected food (e.g., it might kill raw oysters, reducing shelf life).Another class of disease unlikely to be affected by irradiation is bovine spongiform encepalopathy ("mad cow disease"). Although the cause of the disease is still under study, scientists currently believe the culprit to be prion particles, simple collections of protein lacking DNA. Their simple structure makes them resistant to disruption by radiation.

Labeling Requirements

Since food irradiation is classified as a food additive, its presence must be disclosed on a label as per regulations. According to those regulations, the product must bear a legend saying either "Treated with radiation" or "Treated by irradiation." The legend must be accompanied by a symbol known as a radura, an international symbol designed specifically to indicate food irradiation.

Scientific Concerns about Irradiaiton

Although food irradiation is an effective tool, there are legitimate scientific arguments that can be made against food irradiation. Even though the arguments have been substantially refuted, they are still often recited by those unfamiliar with the science of food irradiation.

Nutrition Effects of Irradiation

Macronutrients (proteins, fats and carbohydrates) and minerals (e.g., iron, phosporous and calcium) are substantially unaffected by radiation doses at approved levels. Some vitamins, particularly thiamine, undergo an appreciable reduction when exposed to radiation. The maximal use of food irradiation on meat would result in a decline in the amount of B vitamins consumed in the average person's diet. Irradiation under approved conditions has been demonstrated to have no dangerous effects on food, either chemical or microbial in nature. Although irradiation does reduce nonpathogenic spoilage bacteria (thereby increasing shelf life), the population of spoilage bacteria still exceeds that of pathogenic bacteria, so that the ordinary characteristics of spoiled food will still been present before the food has reached a dangerous state. Irradiation can have undesired side effects on sensory qualities of the food, but such effects tend to be minimal, especially when manufacturers take conscious steps to avoid them.

Food Extrusion

Food extrusion is a form of extrusion used in food processing. It is a process by which a set of mixed ingredients are forced through an opening in a perforated plate or die with a design specific to the food, and is then cut to a specified size by blades. The machine which forces the mix through the die is an extruder, and the mix is known as the extrudate. The extruder consists of a large, rotating screw tightly fitting within a stationary barrel, at the end of which is the die.

Extrusion enables mass production of food via a continuous, efficient system that ensures uniformity of the final product. Food products manufactured using extrusion usually have a high starch content. These include some pasta, breads (croutons, bread sticks, and flat breads), many breakfast cereals and ready-to-eat snacks, confectionery, pre-made cookie dough, some baby foods, full-fat soy, textured vegetable protein, some beverages, and dry and semi-moist pet foods.

Types of Extrusion

There are two types of extrusion processes:

- Cold extrusion, which mixes and shapes foods such as biscuit dough and pasta without cooking them.
- Hot extrusion (or extrusion cooking), which is used to produce a wide range of products, including crisp snack foods, sugar confectionery and soya-based weaning foods.

Both use equipment known as an 'extruder'. This equipment has a screw inside a barrel that conveys materials along the barrel and kneads the food into a semi-solid, plasticized mass. In cold extruders the material is not heated but simply formed into shapes (including rods, tubes, strips or shells), when it is forced through openings in a 'die' at the discharge end of the barrel. In extrudercookers the material is heated by friction and/or supplementary heaters in the barrel and it emerges from the dies under pressure. Some snack food products expand rapidly and have a light, crisp texture, caused by steam being flashed off due to the sudden pressure drop when they emerge from the die. Cold-extruded products are preserved by chilling, baking or drying, whereas extrusion cooking destroys contaminating micro-organisms and the dry products have a long shelf life. They are packaged to prevent them picking up moisture and to prevent oxidation during storage. Cold extruders are suitable for all scales of operation from household- to small-scale, but extrudercookers are much more expensive and are likely to only be affordable by larger-scale producers.

Cold Extrusion

The main application of cold extruders is in pasta production, although similar machines are used to form biscuit dough into different shapes. A pasta extruder is used to make many different types of pasta using dough made from durum wheat flour (or 'semolina') and eggs (optional). Coloured pasta can also be made by adding tomato purée or spinach paste etc.

Different sizes of equipment are available, from small manual machines that are used in microscale production and food service outlets, to larger electric machines. The equipment has a mixing chamber, extruder barrel and a die for the desired pasta shape. There is debate over the best material for dies, with some equipment manufacturers using stainless steel or plastic, and others preferring bronze dies, which they claim imparts a rough surface to the pasta that holds sauce better than pasta made by other methods. Shapes are cut to the appropriate length as they emerge from the die, except rigatoni, which is extruded in long lengths and then cut to the correct size (straight for rigatoni or angled for penne rigati). Pasta is cooked immediately in food service outlets, or dried by processors for retail sale. It can also be frozen for up to six months.

Hot Extrusion

Extruder-cookers may be single- or twin-screw machines. Twin-screw machines have approximately twice the capital and maintenance costs of single screw machines and are unlikely to be affordable by most small-scale processors as compared to single-screw extruders. The two factors that control the type and quality of foods that are produced by hot extrusion are the operating conditions in the extruder barrel and the mixture of ingredients that is used.

Operating Conditions and Process of Extrusion

The important operating conditions are the temperature and pressure in the barrel, the diameter of the die apertures and the product shear rate. The shear rate is influenced by the speed and geometry of the screw (size, number, pitch and diameter of the flights), and by the internal design of the barrel, including grooves in the barrel, or restrictions (known variously as 'throttle rings', 'kneading discs', or 'shearlocks'). Additional heating may be provided by a steam-jacketed barrel, a steam-heated screw, or electric heating elements around the barrel.

'High-shear' extruders have high screw speeds and shallow flights to create the high pressures and temperatures needed to make expanded products; 'medium-shear' extruders are used to make texturised proteins and semi-moist foods; and 'low shear' extruders have a deep-flighted screw that operates at low speeds in a smooth barrel to create low pressures for forming meat products or gums. The selection of the correct type of extruder for a particular application should take account of the types of ingredients and the properties required in the product (e.g. its bulk density, texture, colour and other sensory properties) and the required production rate.

Dies have different shaped holes (e.g. round holes to produce rods, square holes for bars, or slots to produce sheets), or they may produce more complex shapes. The temperature and moisture content of the food and the extent of shearing in the barrel control the amount of expansion of the product and hence its texture. Some products require the dies to be heated to give the required degree of expansion, whereas others have cooled dies to reduce expansion. There are therefore a very large number of potential combinations of equipment design features and small-scale processors should seek advice from extruder manufacturers before purchasing a machine.

In the extrusion process, raw materials are first ground to the correct particle size, usually the consistency of coarse flour. The dry mix is passed through a pre-conditioner, in which other ingredients are added depending on the target product; these may be liquid sugar, fats,dyes, meats or water. Steam is injected to start the cooking process, and the preconditioned mix (extrudate) is then passed through an extruder. The extruder consists of a large, rotating screw tightly fitting within a stationary barrel, at the end of which is the die. The extruder's rotating screw forces the extrudate toward the die, through which it then passes. The amount of time the extrudate is in the extruder is the residence time.

The extruded product usually puffs and changes texture as it is extruded because of the reduction of forces and release of moisture and heat. The extent to which it does so is known as the expansion ratio. The extrudate is cut to the desired length by blades at the output of the extruder, which rotate about the die openings at a specific speed. The product is then cooled and dried, becoming rigid while maintaining porosity.

The cooking process takes place within the extruder where the product produces its own friction and heat due to the pressure generated (10–20 bar). The process can induce both protein denaturation and starch gelatinization, depending on inputs and parameters.

Many food extrusion processes involve a high temperature over a short time. Important factors of the extrusion process are the composition of the extrudate, screw length and rotating speed, barrel temperature and moisture, die shape, and rotating speed of the blades. These are controlled based on the desired product to ensure uniformity of the output.

Moisture is the most important of these factors, and affects the mix viscosity, acting to plasticize the extrudate. Increasing moisture will decrease viscosity, torque, and product temperature, and increase bulk density. This will also reduce the pressure at the die. Most extrusion processes for food processing maintain a moisture level below 40%, that is low to intermediate moisture. Highmoisture extrusion is known as wet extrusion, but it was not used much before the introduction of twin screw extruders (TSE), which have a more efficient conveying capability. The most important rheological factor in the wet extrusion of high-starch extrudate is temperature.

The amount of salt in the extrudate may determine the colour and texture of some extruded products. The expansion ratio and airiness of the product depend on the salt concentration in the extrudate, possibly as a result of a chemical reaction between the salt and the starches in the extrudate. Colour changes as a result of salt concentration may be caused by "the ability of salt to change the water activity of the extrudate and thus change the rate of browning reactions". Salt is also used to distribute minor ingredients, such as food colours and flavours, after extrusion; these are more evenly distributed over the product's surface after being mixed with salt.

Ingredient Mixture

Different mixtures of ingredients produce completely different products when the same operating conditions are used in the same extruder. This is because starch, proteins, moisture and other ingredients (e.g. oil or an emulsifier) have different effects on the structure and texture of the extruded food. For example, starch or proteins create a three-dimensional structure that contains the other ingredients. Starches from cereal or legume flours (e.g. maize, wheat, rice, barley, pea, bean), or from tuber flours (e.g. potato, cassava, tapioca) are used for extruded breakfast cereals, snack foods, pasta and biscuits. Proteins from soybeans, sunflower seeds, rapeseed, or gluten from wheat, are used to make meat-like products such as texturised vegetable protein.

When making extruded snack foods from cereal or potato starch, the process operates at high temperatures (130 – 180 °C) to produce a 'fluid melt' that contains gelatinised starch and superheated water vapour. When this leaves the extruder it expands to form hard, porous, brittle snack food products. Flavourings and/or colourings are sprayed onto the food after it is extruded. Snack foods can also be made from extruded 'preforms' or 'half-products'. These are small, dense extruded pellets that are sold to other processors to make the final snack food by frying or toasting the pellets. When half-products are heated, they soften and the residual moisture in the pellets turns to steam, which rapidly expands the pellets to produce the snack food.

Extruded weaning foods are produced as flakes or pellets from a mixture of cereal and legume flours that have the correct protein and energy content for growing children. The extruded products may also be fortified with minerals and vitamins. In use, the products are ground to a powder and mixed with hot water to form a porridge that is fed to children. The high temperatures used in the extruder ensure that products are safe and have a shelf life in excess of 12 months when packed in moisture proof and airtight packaging. The process is used for both commercial weaning foods and foods used as emergency or supplementary foods by development agencies.

Extrusion cooking is also used to produce sugar confectionery products such as liquorice, toffee, fudge and boiled sweets from sugar, glucose and starch. Hard-boiled sweets are produced from sugar and corn syrup with added colours, acids and flavours. They are extruded as a 'rope' and made into the required shapes using stamping or forming machines. Energy consumption using an extruder to produce confectionery is about half of that used in boiling pans.

Extruded Products

Extrusion has enabled the production of new processed food products and "revolutionized many conventional snack manufacturing processes". The various types of food products manufactured by extrusion typically have a high starch content. Directly expanded types include breakfast cereals and corn curls, and are made in high temperature, low moisture conditions under high shear. Unexpanded products include pasta, which is produced at intermediate moisture (about 40%) and low temperature. Texturized products include meat analogues, which are made using plant proteins ("textured vegetable protein") and a long die to "impart a fibrous, meat-like structure to the extrudate", and fish paste. Confectionery made via extrusion includes chewing gum, liquorice, and toffee.

Some processed cheeses and cheese analogues are also made by extrusion. Processed cheeses extruded with low moisture and temperature "might be better suited for manufacturing using extrusion technology" than those at high moisture or temperature. Lower moisture cheeses are firmer and chewier, and cheddar cheese with low moisture and an extrusion temperature of 80 °C was preferred by subjects in a study to other extruded cheddar cheese produced under different conditions. An extrudate mean residence time of about 100 seconds can produce "processed cheeses or cheese analogues of varying texture (spreadable to sliceable)".

Other food products often produced by extrusion include some breads (croutons, bread sticks, and flat breads), various ready-to-eat snacks, pre-made cookie dough, some baby foods, some beverages, and dry and semi-moist pet foods. Specific examples include macaroni, jelly beans, sevai, and some french fries. Extrusion is also used to modify starch and to pellet animal feed.

Effects of Extrusion on Nutritional Quality

Extrusion enables mass production of food via a continuous, efficient system that ensures uniformity of the final product. This is achieved by controlling various aspects of the extrusion process. It has also enabled the production of new processed food products and "revolutionized many conventional snack manufacturing processes". The extrusion process results in "chemical reactions that occur within the extruder barrel and at the die". Extrusion has the following effects:

- Destruction of certain naturally occurring toxins.
- Reduction of microorganisms in the final product.
- Slight increase of iron-bioavailability.
- Creation of insulin-desensitizing starches (a potential risk-factor for developing diabetes).
- Loss of lysine, an essential amino acid necessary for developmental growth and nitrogen management.
- Simplification of complex starches, increasing rates of tooth decay.
- Increase of glycemic index of the processed food, as the "extrusion process significantly increased the availability of carbohydrates for digestion".
- Destruction of Vitamin A (beta-carotene).
- Denaturation of proteins.

The material of which an extrusion die is made can affect the final product. Compared to stainless steel dies, a pasta machine with bronze dies produces a rougher surface. This is considered to give an improved taste, as it better retains pasta sauces. "Bronze die" pasta is labelled as such on retail packages, to indicate a premium product.

The effects of "extrusion cooking on nutritional quality are ambiguous", as extrusion may change carbohydrates, dietary fibre, the protein and amino acid profile, vitamins, and mineral content of the extrudate in a manner that is beneficial or harmful. High-temperature extrusion for a short duration "minimizes losses in vitamins and amino acids". Extrusion enables mass production of some food, and will "denature anti-nutritional factors", such as destroying toxins or killing microorganisms. It may also improve "protein quality and digestibility", and affects the product's shape, texture, colour, and flavour. It may also cause the fragmentation of proteins, starches, and non-starch polysaccharides to create "reactive molecules that may form new linkages not found in nature". This includes Maillard reactions which reduce the nutritional value of the proteins. Vitamins with heat lability may be destroyed. Nutritional quality has been found to improve with moderate conditions (short duration, high moisture, low temperature), whereas a negative effect on nutritional quality of the extrudate occurs with a high temperature (at least 200 °C), low moisture (less than 15%), or improper components in the mix.

Vacuum Packing

Vacuum packing is a packaging process that preserves food by creating a vacuum that removes the air before sealing, preventing the growth of micro-organisms. Food can then be stored for longer, though vacuum packing is not a substitute for refrigeration or freezing.

It is necessary to take adequate precautions when vacuum packing. Spoilage bacteria is less likely to grow with the removal of oxygen. However, pathogenic bacteria (which causes illness) may have opportunity to reproduce and the changes they make to the food may not be detectable by sight or smell. All vacuum packaged foods should be clearly labelled and dated to manage the preservation process.

Equipment is required for vacuum packing food. Domestic vacuum sealers to vacuum package food at home are available in Ireland at a cost of approximately €40. Periodically they are available from multiple stores, and can also be found online.

Keep the following food safety guidelines in mind when vacuum packing food:

- Wash your hands before and during the vacuum packing process.
- Try not to touch food with your bare hands.
- Keep all utensils, bags, and work surfaces clean.
- Label each package.
- Remember that foods which require refrigeration/freezing ordinarily, must also be kept in the fridge/freezer once vacuum packed.

If you have been wondering whether a vacuum sealing machine would be of use in your life, here are some benefits to owning a food vacuum sealer:

- Vacuum Food Sealers Save your Time: Seal family-sized portions of meals or individual portions of snacks or vegetables and freeze. When it's time to eat, fix the portions you need and you're done. This is especially useful when cooking sous vide – pop the portions you need into your VacMaster sous vide pot and cook until done. Dinner is served.
- Vacuum Sealers Save your Money: Buying meats in bulk, dividing into meal-sized portions and sealing allows you to have fresh meat on hand for up to 6 months. This leads to an incredible cost savings over buying your meat from the counter before every meal.
- Vacuum Sealers Preserve Food: Vacuum sealers preserve food by preventing the growth of mold or bacteria. Vacuum sealing deprives your food of oxygen. Mold and bacteria cannot grow without oxygen.

- Vacuum Sealing Protects Food: Vacuum sealing protects food from dehydration and freezer burn. Freezer burn occurs when water gets into your food. The vacuum bag keeps your food from contact with the air, so moisture cannot evaporate, making it an excellent barrier from the elements.

- Vacuum Sealing Extends Shelf Life: The shelf life of your food is greatly extended when you vacuum seal it. Vacuum sealing food preserves it 3 to 5 times longer than food stored in plastic containers or bags. The exact amount of time food will last in vacuum sealing varies depending on whether you are storing in the refrigerator, freezer or pantry.

- Food Lasts Longer: Frozen food that is vacuum sealed lasts an average of 2-3 years, while it will last 6-12 months, on average, stored in other ways. Most vacuum sealed foods will last in the refrigerator for 1-2 weeks, which is much longer than the typical 1-3 days food will last when stored conventionally in a refrigerator.

- Efficient Food Organization: Vacuum sealing makes for efficient, organized packaging. Vacuum sealed food takes up less room in your refrigerator or freezer and allows you to easily see the foods you are storing.

- No Freezer Burn with Chamber Vacuum Sealers: Vacuum sealing provides food an air-tight environment, preventing the crystals that cause freezer burn from forming on your food. Vacuum sealing keeps the moisture in your food for an extended period of time, allowing you to enjoy food that tastes fresh months after you have stored it.

- Vacuum Sealed Food Tastes Better: Vacuum sealing holds in the moisture, juices, and flavor of the food. You'll notice the difference in taste.

- Vacuum Sealed Keeps Food Safe: Vacuum sealing removes most of the oxygen, preventing chemical reactions triggered by oxygen that cause food to spoil. It also prevents bacteria buildup that can spoilage in foods.

- Vacuum Sealed Keeps Food Fresh: Vacuum sealing not only keeps food from spoiling, it keeps fats in the food from going rancid, keeps foods from changing color as they often do in the freezer, and prevents them from getting a slimy film or bad odor.

- Store Liquids Easy Vacuum Sealed: Place the liquid in the freezer for a couple hours to harden it a little. Then place the partially frozen liquid into a sealing bag, leaving room at the top for sealing. You can also seal liquids inside a jar and vacuum seal the jar.

- Vacuum Seal Leftovers: Vacuum seal any leftover food from your meal right away. This eliminates you having to throw out the food and gives you another easy meal for the future. Freeze the leftover portions and they will last for months – keeping you from having to eat the same meal three nights in a row. Pull it out a few weeks later for a meal that tastes fresh.

- Reseal Chips: A vacuum food sealer is handy for resealing chips, just remember not to use the vacuum suction or you will crush the contents. Allow several inches of open space at the top of the bag, press the seal button and insert the bag (ensuring the top is in straight) in the sealer.

- Vacuum Seal Trail Mix and Snacks: Separate snacks or trail mix into individual portions and seal. This makes it easy to take snack-sized servings with you when on the go – keep them in your purse, backpack, or lunch box for a readily available energy boost.

To further extend the life of your food, thoroughly clean it before storing. Also, between packaging wipe down any surfaces used to prevent cross contamination. VacMaster offers vacuum sealing machines and pouches to get you organized while preserving your food.

Food Concentration

The heat is used as the most common method of liquid food concentration. Water activity of the liquid food is reduced due to the loss of moisture by evaporation. But, this method of liquid food concentration results in loss of sensory and nutritional properties of the food. Hence there is a need for the alternative method of liquid food concentration.

Freeze concentration has been considered as an efficient method. In this method, water is removed as ice rather than vapour. In freeze concentration the product is cooled below its freezing point allowing water to form crystals which are then removed from the concentrate by centrifugation. Freeze concentration selectively removes water from the food without altering any other components. The preservation effect of freeze concentration is due to the reduction in the water activity, without the application of heat. The low temperatures used in the process cause a high retention of volatile aroma compounds, as a result of which nutritional qualities and sensory characteristics are better maintained. Therefore, the concentrated product gets higher quality than that processed by thermal evaporation.

Steps Involved in Freeze Concentration

When a solid or liquid food is frozen, the components present in the food will not freeze at once. First the water gets frozen forming ice crystals in the mixture. The remaining unfrozen food solution is now higher in solids concentration. Before the entire mixture freezes, it is possible to separate the initially formed ice crystals. This separation could be done using centrifugal separation or filter press. The concentrated unfrozen food solution is again passed through the screen while the frozen water crystals are retained and can be discarded. Repeating this process several times on the concentrated unfrozen food solution can increase its final concentration several-fold.

Freeze Concentration

The processes involved in freeze concentration are:

- Fractional crystallization of water to ice,
- Growth of ice crystals,
- Removal of the ice.

Fractional crystallization of water to ice: It done using direct contact freezing system (for example solid carbon dioxide or liquid nitrogen) or indirect equipment to freeze the liquid food.

Growth of Ice Crystals

The ice crystals formed are slowly agitated by a mixing vessel, such that it grows to bigger size. A larger crystal generally gets separated from the concentrate easily. In freeze concentration, it is preferred to grow the ice crystal bigger in order to reduce the amount of concentrated liquid entrained with the crystals. It is done by slowly stirring the slurry (mixture of ice crystals and liquid) and permitting the large crystals to grow.

Separation of Crystals from the Concentrated Solution

Separation of the crystal so formed due to the conversion of water to ice is achieved either using gravity drainage, filtration, filter pressing, centrifugation or wash columns. Wash column - functions by feeding the ice-concentrate slurry into the bottom of a vertical cylinder. The greater part of the concentrate passes through the crystals and is separated. Heater is placed at the top of the column, to melt the formed ice crystals, and the portion of the melted water passes through the bed of ice crystals to remove entrained concentrate.

Concentration is done in single-stage or, more, generally multi-stage equipments are used. A multi-stage concentrator system consumes less energy and has higher production rates. Advancement in techniques for generating bigger ice crystals and more efficient washing has increased the maximum obtainable concentration to 45 % solids. The efficiency of separation of the crystal from the concentrated liquor is determined by the degree of clumping of the crystals, and amount of liquor entrained.

Industrial Applications of Freeze Concentration

- The application of freeze concentration includes concentration of orange juice, raise the alcohol content of wine, pre-concentrating coffee and tea extract prior to freeze drying.
- Helps in the concentration of milk products, fruit juice, beer and other flavor products.
- Among all the other application, concentration of alcoholic beverage is the one in which freeze concentration is superior to other methods.

Advantages:

- This process can be carried out at low temperature.
- High retention of volatile aroma.

Disadvantages:

- The concentration achieved by freeze concentration is (40–45 °Brix) lower than the values obtained by evaporation (60–65 °Brix).
- Separation of smaller ice crystals formed during rapid freezing is difficult.

- Freeze concentration is not suitable for highly pulped liquids.
- High refrigeration cost, high equipment investments to handle frozen mass and high operational costs.
- Freeze concentration has low production rate than conventional dehydration, evaporation or membrane concentration. Due to these reasons freeze concentration is used only for high value product.

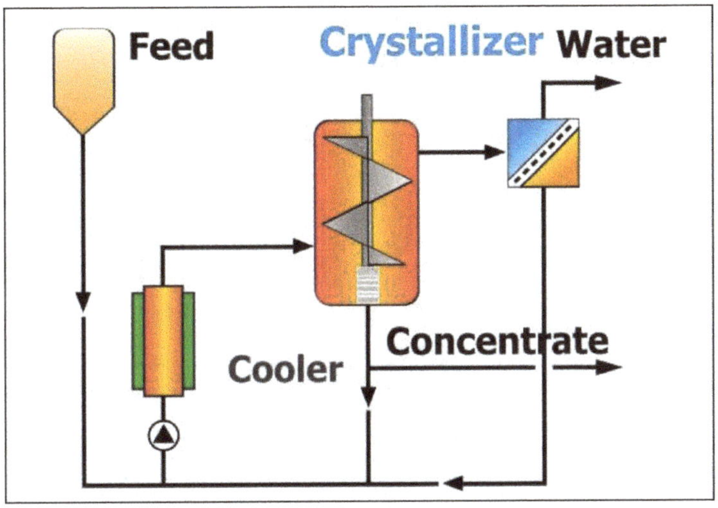

Concentration by Membranes

Evaporative concentration of fluid food has the following disadvantages:

- Heat induced deterioration of sensory (colour, taste, aroma) properties.
- Deterioration of nutritional value (vitamins) of finished products (concentrate).

It is well known fact that, most of the aroma compounds present in the liquid food products like raw juice are lost and the aroma profile undergoes irreversible changes in the first few minutes of evaporative concentration process.

Concentration by Membrane Separation Processes

In membrane separation process, a membrane is placed between two phases (feed and permeate). Membranes are materials that have voids in them, allowing some molecules to pass more conveniently than some other molecules. A semi-permeable membrane is a very thin film that allows some types of matter to pass through while leaving others behind.

Membrane separation system, a fluid containing two or more components is in contact with a membrane that permits selected components (for example, water in the fluid) to permeate more readily than other components. One of the earliest successful industrial applications of membrane technology was in the food industry. In 1972, a dairy plant in New York began processing cheese whey by reverse osmosis. Some of the application of the membrane separation process in food industry includes purification of water, concentration and clarification of fruit juices, milk products, alcoholic beverages, and this technique can be used in treatment of food industry waste.

In order to understand the principle behind the membrane based separation we need to understand the principles of osmosis. Osmosis is a naturally occurring phenomenon and one of the most important processes in nature. It is a process where a lower concentrated solution will tend to migrate to a strong concentrated solution (as shown in the figure), same the way plant root absorbs water from soil. Once the concentration in both chambers have reached equilibrium, the movement stops as enough pressure has built up to stop the flow of water from one chamber to the other. This pressure is known as osmotic pressure.

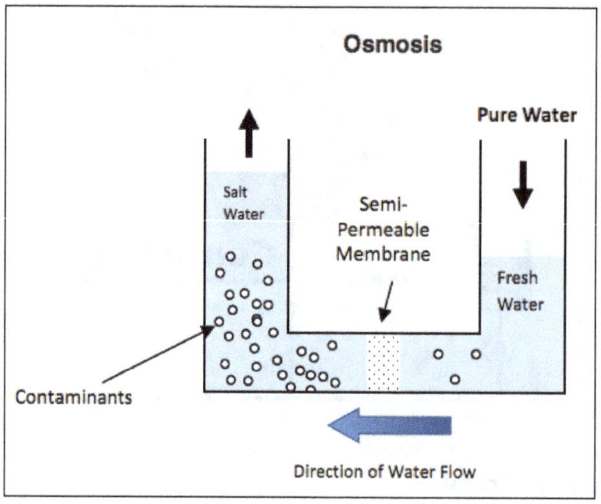

A semipermeable membrane allows only some atoms or molecules to pass through. The sizes of the pores are such that, it will allow water vapour to pass through, preventing liquid water from passing.

Reverse Osmosis

Osmosis occurs naturally. Energy need to be supplied to reverse the process of osmosis. Reverse osmosis is a process in which pressure greater than the natural osmotic pressure is applied on the high concentration side of the membrane, forcing the water to travel through the membrane from the higher concentration to low concentration, thus 'reversing' the natural tendency of water flow. Reverse osmosis required high pumping pressures, to push the water through the membrane.

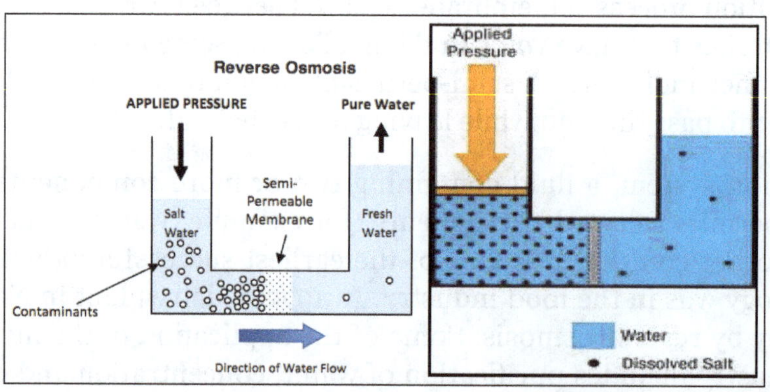

Membrane separations are classified by pore size as Microfiltration (MF), Ultrafiltration (UF), Nanofiltration (NF), Reverse Osmosis (RO) and Forward Osmosis (FO).

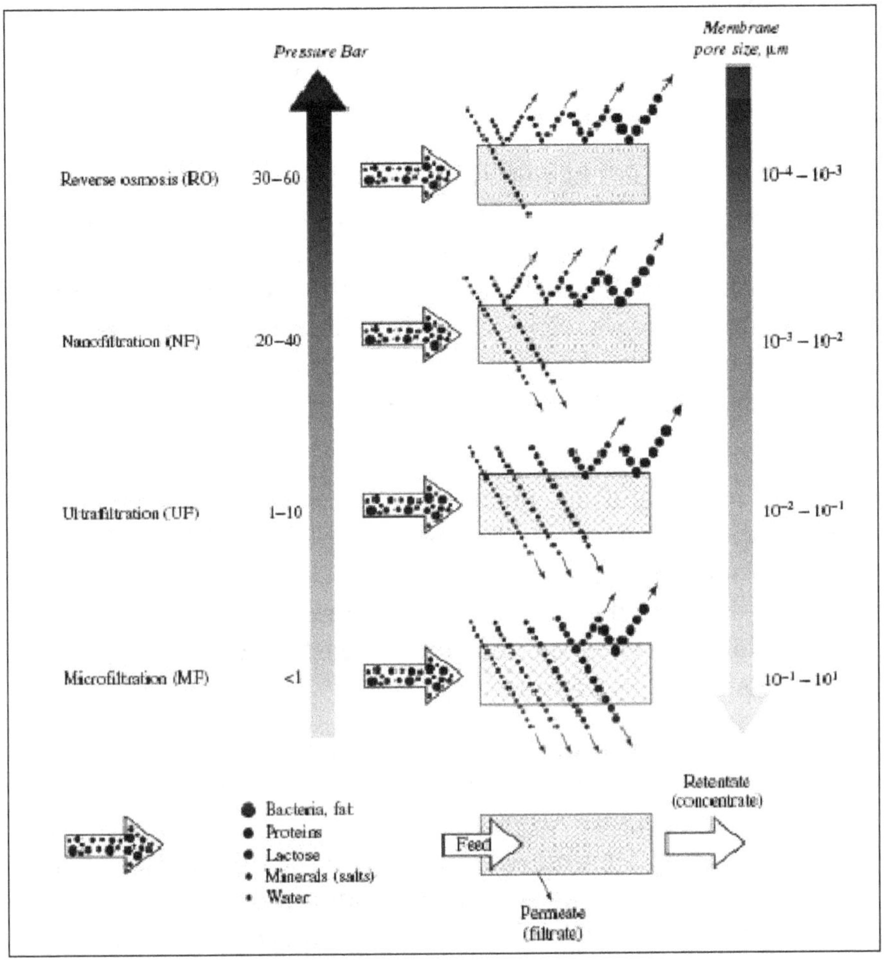

The figure illustrates the spectrum of application of membrane separation processes in the food industry. There is no sharp division between the processes; membrane pore size and the applied-pressure difference are often used to distinguish the processes from each other.

The only problem in membrane separation is that during this concentration process on the retentate side of the membrane, the concentration of the total soluble solids becomes higher and higher, causing higher osmotic pressure and decreasing the driving force of the concentration process, which is the difference between the transmembrane pressure and osmotic pressure of the retentate (fruit juice concentrate) and permeate (practically water).

Ultrafiltration

Ultrafiltration membranes have a higher porosity and retain only large molecules (for example proteins or colloids) which have a lower osmotic pressure. Smaller solutes are transported across the membrane with the water. Ultrafiltration membranes have a porosity of 1-100nm. Ultrafiltration therefore operates at lower pressure (0.01 to 0.055 MPa). Ultrafiltration can be carried out at room temperature or in a cold room. Main application of ultrafiltration in food industry is the concentration of milk, recovery of whey proteins, recovery of potato starch and proteins, concentration of egg products, clarification of fruit juices and alcoholic beverages. Ultrasound technique is also used in pharmaceutical industry, chemical industry, textile industry and paper industry.

Micro filtration (MF)

Micro filtration (MF) is the method of removing particles (0.025 µm to 10.0µm) from the fluids by allowing it to pass through a microporous medium such as a membrane filter. In this, membranes have pore diameter from 0.1 to 10 µm, the hydrostatic pressure differences used are in the range of 0.01–0.1 MPa. From the outside of the membrane the liquid is pumped inn, and from the inside of membrane the clean water is collected. A filter cake that develops on the membrane is needed to be removed by back washing. Recovery rates are more than 90%, with feed sources having low turbidity. This method is useful for harvesting microbial cells from fermentation broths.

Membranes can be used for separating constituents of foods on a molecular basis and thus for a membrane to work efficiently we require to understand all the ideal conditions required for a membrane. The optimal conditions for a membrane are it should offer high degree of separation, it should withstand high pressure, and it should be operational at moderate temperatures and at low pH ranges, resist microbiological effects and should be cost effective. Majority of the membranes are made up of Cellulose Acetate, Polyamides.

Polysulfones, Polyacrylonitrile, Polypropylene and Inorganic Ceramic. The operational temperature ranges for Cellulose Acetate (T<30 °C), Polysulfones and polyamides (T<80 °C), Ceramic (T<130 °C).The pH range for Cellulose Acetate is (3-8) and for Polysulfones it is (3- 11).The average thickness should be 0.5-1µm. Membrane modules used in microfiltration are Plate and frame module used in filtering dilute solutions and Spiral-wound module which is preferred when the solutions handled are heavily concentrated and in conditions of high temperatures and extreme pH.

Nano-filtration

Nano-filtration membranes have pore size of approximately 0.001 microns. Nano-filtraiton requires high operating pressure than either microfiltration or ultrafiltration. These systems can remove virtually all cysts, bacteria, viruses and humic materials. There are two types of membranes used in separation. Spiral membranes which is more reactive to pollution and tubular membranes which is not easily prone to pollution. Some of the applications of nano-filtration include colour removal, removal of low molecular weight species such as organics in pharmaceutical production, surfactants from waste waters. In food industries, nano-filters are used in demineralization of whey, fruit juice concentration, molasses purification for reuse and in waste water treatments.

Forward Osmosis

Forward osmosis (FO), also known as direct osmosis, is a promising membrane technology and its only driven force is a difference in osmotic pressure. The transfer of water occurs from low concentration side to high concentrated side across a semi-permeable membrane, until osmotic pressure differences between sides are nil. Osmotic pressure difference is the only driving force for the movement of water.

FO Advantages

- Low hydraulic pressure for operation, which reduces the cost for electrical energy.
- Low processing temperature, which avoids the thermal degradation of food quality factors and conserves electrical energy.
- High product recovery and low discharge of brine to the environment.
- Low irreversible fouling, which leads to low costs on membrane cleaning and replacement.
- Efficient treatment of feeds with a high solids content.
- Modularity and ease of scale-up.

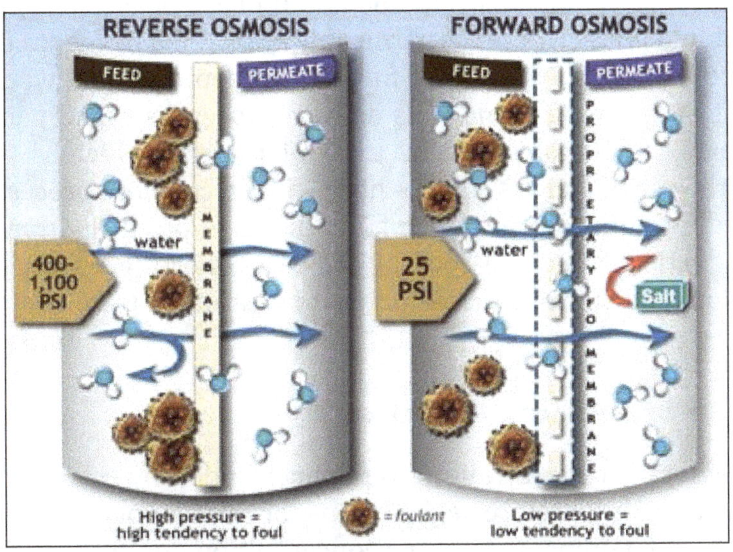

Membrane Distillation for Concentration

Membrane distillation is a membrane technique that enables separation due phase change, and is achieved using thermal driving force. A hydrophobic membrane exhibits a barrier for the liquid phase and allows only the vapour phase to pass through the membrane pores. The drive force for this process is provided by partial vapour pressure difference caused by temperature difference.

The surfaces of the membrane are in direct contact with two liquid phases, the feed (warm solution) and the permeate (cold solution), kept at different temperatures. Vapour is transported, through the membrane as gaseous phase exist inside the membrane, and the movement of the vapor continues, until the pressure difference is maintained. The vapor diffuses through the membrane pores and condenses at the permeate membrane interface. Thus, membrane distillation depends on vapor liquid equilibrium as a basis for separation, and to achieve the characteristics phase change, latent heat of vaporization need to be supplied.

Vacuum Membrane Distillation: Vacuum membrane distillation is a membrane separation process that removes volatile compounds from the aqueous solution. In this one side of the membranes is passed with the feed whereas the other side is vacuumed. The membrane is hydrophobic and micro porous and so the solution in the feed side do not pass through the membrane towards the vacuum side due to interface tension. When the pressure of the vacuum side is kept below the equilibrium vapor pressure of the solution, the water present in feed side gets evaporated and move towards the vacuum side. Meanwhile, the vapor entering the vacuum side is vacuumed outside the membrane cell.

Dialysis (Liquid Permeation)

Dialysis is a membrane process where dissolved substance that has a molecular weight less than 100 Dalton diffuses from one side of the membrane (feed side) to the other (permeate side) according to its concentration gradient. The solutes separate by passing through the membrane that acts as a fibre filter and the separation occurs by a screening action based on the pore diameter and particle size. In this case the small solutes in one liquid phase penetrate through a porous membrane to the another liquid phase where the permeants are diluted by means of a sweeping solvent. The permeants can be separated by flushing from the sweeping liquid when its boiling point is much lower than that of sweeping liquid, this dialysis process is called pertraction. In practice, dialysis is used in separating microbial species that differ in size, which have a large difference in diffusion rates. With the use of dialysis sodium hydroxide in cellulose processing can be removed, acids can be recovered from metallurgical liquors, removal of products from a culture solution in fermentation, and reduction of alcohol content of beer. Other therapeutic applications include dialysis in removing toxic substances such as urea and creatinine, from the blood of patients with chronic uremia.

Advantages of Membrane Concentration

The advantages of membrane concentration over concentration by evaporation are primarily the food is not heated, and there is therefore negligible loss of volatiles or changes to nutritional or eating quality. In contrast with boiling, membrane concentration does not involve a change in phase and therefore uses energy more efficiently. In addition it's simple installation with lower labour and operating costs and requires no requirement for steam boilers. While the main limitations of membrane concentration are: variation in the product flow rate when changes occur in the concentration of feed liquor, higher capital costs than evaporation a maximum concentration to 30% total solids, fouling of the membranes (deposition of polymers), which reduces the operating time between membrane cleaning.

Industrial Applications of Membrane Concentration

- Milk fat fractionation from whole milk.
- Removal of bacteria and spores from skimmed milk.
- Production of milk protein and native casein concentrates recovery.
- Fractionation of whey proteins.

- Early application of membrane technology in the dairy industry was in cheese manufacturing and brine treatment by UF.
- Beer Dealcoholisation.
- Beer Clarification.
- A diafiltration process by Reverse osmosis is applied to enhance the wine by removing the negative aroma components causing the stale taste with the permeate.
- The combination of RO and evaporation can also provide an interesting process for the concentration of apple juice.
- Membrane Processes for Water and Wastewater.

Chapter 5

Food Contamination: Symptoms and Effects

The corruption of food by a physical, biological or chemical substance is termed as food contamination. It can cause food allergy as well as foodborne illnesses such as diarrhea, gastroenteritis, nausea and vomiting. This chapter discusses in detail these effects of food contamination.

Foodborne Illness

Foodborne illness (also foodborne disease and colloquially referred to as food poisoning) is any illness resulting from the food spoilage of contaminated food, pathogenic bacteria, viruses, or parasites that contaminate food, as well as chemical or natural toxins such as poisonous mushrooms and various species of beans that have not been boiled for at least 10 minutes.

Symptoms vary depending on the cause, and are described below in this article. A few broad generalizations can be made, e.g.: The incubation period ranges from hours to days, depending on the cause and on how much was consumed. The incubation period tends to cause sufferers to not associate the symptoms with the item consumed, and so to cause sufferers to attribute the symptoms to gastroenteritis for example. Symptoms often include vomiting, fever, and aches, and may include diarrhea. Bouts of vomiting can be repeated with an extended delay in between, because even if infected food was eliminated from the stomach in the first bout, microbes (if applicable) can pass through the stomach into the intestine via cells lining the intestinal walls and begin to multiply. Some types of microbes stay in the intestine, some produce a toxin that is absorbed into the bloodstream, and some can directly invade deeper body tissues.

Causes

Foodborne illness usually arises from improper handling, preparation, or food storage. Good hygiene practices before, during, and after food preparation can reduce the chances of contracting an illness. There is a consensus in the public health community that regular hand-washing is one of the most effective defenses against the spread of foodborne illness. The action of monitoring food to ensure that it will not cause foodborne illness is known as food safety. Foodborne disease can also be caused by a large variety of toxins that affect the environment.

Foodborne illness can also be caused by pesticides or medicines in food and naturally toxic substances such as poisonous mushrooms or reef fish.

Poorly stored food in a refrigerator.

Bacteria

Bacteria are a common cause of foodborne illness. In the United Kingdom during 2000, the individual bacteria involved were the following: *Campylobacter jejuni* 77.3%, *Salmonella* 20.9%, *Escherichia coli* O157:H7 1.4%, and all others less than 0.56%. In the past, bacterial infections were thought to be more prevalent because few places had the capability to test for norovirus and no active surveillance was being done for this particular agent. Toxins from bacterial infections are delayed because the bacteria need time to multiply. As a result, symptoms associated with intoxication are usually not seen until 12–72 hours or more after eating contaminated food. However, in some cases, such as Staphylococcal food poisoning, the onset of illness can be as soon as 30 minutes after ingesting contaminated food.

Most common bacterial foodborne pathogens are:

- *Campylobacter jejuni* which can lead to secondary Guillain–Barré syndrome and periodontitis.
- *Clostridium perfringens*, the "cafeteria germ".
- *Salmonella* spp. – its *S. typhimurium* infection is caused by consumption of eggs or poultry that are not adequately cooked or by other interactive human-animal pathogens.

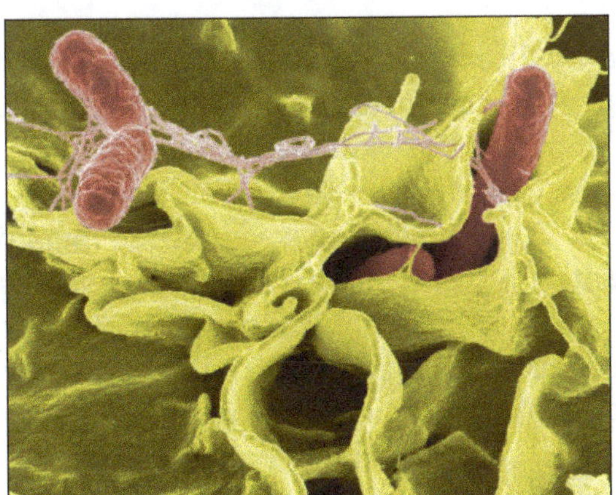

Salmonella

- *Escherichia coli O157:H7* enterohemorrhagic (EHEC) which can cause hemolytic-uremic syndrome.

Other common bacterial foodborne pathogens are:

- *Bacillus cereus.*
- *Escherichia coli*, other virulence properties, such as enteroinvasive (EIEC), enteropathogenic (EPEC), enterotoxigenic (ETEC), enteroaggregative (EAEC or EAgEC).
- *Listeria monocytogenes.*
- *Shigella* spp.
- *Staphylococcus aureus.*
- *Staphylococcal enteritis.*
- *Streptococcus.*
- *Vibrio cholerae*, including O1 and non-O1.
- *Vibrio parahaemolyticus.*
- *Vibrio vulnificus.*
- *Yersinia enterocolitica* and *Yersinia pseudotuberculosis.*

Less common bacterial agents:

- *Brucella* spp.
- *Corynebacterium ulcerans.*
- *Coxiella burnetii* or Q fever.
- *Plesiomonas shigelloides.*

Enterotoxins

In addition to disease caused by direct bacterial infection, some foodborne illnesses are caused by enterotoxins (exotoxins targeting the intestines). Enterotoxins can produce illness even when the microbes that produced them have been killed. Symptom appearance varies with the toxin but may be rapid in onset, as in the case of enterotoxins of *Staphylococcus aureus* in which symptoms appear in one to six hours. This causes intense vomiting including or not including diarrhea (resulting in staphylococcal enteritis), and staphylococcal enterotoxins (most commonly staphylococcal enterotoxin A but also including staphylococcal enterotoxin B) are the most commonly reported enterotoxins although cases of poisoning are likely underestimated. It occurs mainly in cooked and processed foods due to competition with other biota in raw foods, and humans are the main cause of contamination as a substantial percentage of humans are persistent carriers of *S. aureus*. The CDC has estimated about 240,000 cases per year in the United States.

- *Clostridium botulinum.*
- *Clostridium perfringens.*
- *Bacillus cereus.*

The rare but potentially deadly disease botulism occurs when the anaerobic bacterium *Clostridium botulinum* grows in improperly canned low-acid foods and produces botulin, a powerful paralytic toxin.

Pseudoalteromonas tetraodonis, certain species of Pseudomonas and Vibrio, and some other bacteria, produce the lethal tetrodotoxin, which is present in the tissues of some living animal species rather than being a product of decomposition.

Emerging Foodborne Pathogens

Many foodborne illnesses remain poorly understood.

- *Aeromonas hydrophila, Aeromonas caviae, Aeromonas sobria.*

Preventing Bacterial Food Poisoning

Proper storage and refrigeration of food help in the prevention of food poisoning.

Prevention is mainly the role of the state, through the definition of strict rules of hygiene and a public services of veterinary surveying of animal products in the food chain, from farming to the transformation industry and delivery (shops and restaurants). This regulation includes:

- Traceability: in a final product, it must be possible to know the origin of the ingredients (originating farm, identification of the harvesting or of the animal) and where and when it was processed; the origin of the illness can thus be tracked and solved (and possibly penalized), and the final products can be removed from the sale if a problem is detected.
- Enforcement of hygiene procedures such as HACCP and the "cold chain".
- Power of control and of law enforcement of veterinarians.

In August 2006, the United States Food and Drug Administration approved Phage therapy which involves spraying meat with viruses that infect bacteria, and thus preventing infection. This has raised concerns, because without mandatory labelling consumers would not be aware that meat and poultry products have been treated with the spray.

At home, prevention mainly consists of good food safety practices. Many forms of bacterial poisoning can be prevented by cooking it sufficiently, and either eating it quickly or refrigerating it effectively. Many toxins, however, are not destroyed by heat treatment.

Techniques that help prevent food borne illness in the kitchen are hand washing, preventing cross-contamination, proper storage, and maintaining cooking temperatures. In general, freezing or refrigerating prevents virtually all bacteria from growing, and heating food sufficiently kills parasites, viruses, and most bacteria. Bacteria grow most rapidly at the range of temperatures between 40 °F and 140 °F, called the "danger zone". Storing food below or above the "danger zone" can effectively limit the production of toxins. For storing leftovers, the food must be put in shallow containers for quick cooling and must be refrigerated within two hours. When food is reheated, it must reach an internal temperature of 165 °F (74 °C) or until hot or steaming to kill bacteria.

Mycotoxins and Alimentary Mycotoxicoses

The term alimentary mycotoxicoses refers to the effect of poisoning by Mycotoxins (The term 'mycotoxin' is usually reserved for the toxic chemical products produced by fungi that readily colonize crops) through food consumption. Mycotoxins sometimes have important effects on human and animal health. For example, an outbreak which occurred in the UK in 1960 caused the death of 100,000 turkeys which had consumed aflatoxin-contaminated peanut meal. In the USSR in World War II, 5,000 people died due to Alimentary Toxic Aleukia (ALA). The common foodborne Mycotoxins include:

- Aflatoxins – originated from Aspergillus parasiticus and Aspergillus flavus. They are frequently found in tree nuts, peanuts, maize, sorghum and other oilseeds, including corn and cottonseeds. The pronounced forms of Aflatoxins are those of B1, B2, G1, and G2, amongst which Aflatoxin B1 predominantly targets the liver, which will result in necrosis, cirrhosis, and carcinoma. In the US, the acceptable level of total aflatoxins in foods is less than 20 μg/kg, except for Aflatoxin M1 in milk, which should be less than 0.5 μg/kg. The official document can be found at FDA's website.
- Altertoxins – are those of Alternariol (AOH), Alternariol methyl ether (AME), Altenuene (ALT), Altertoxin-1 (ATX-1), Tenuazonic acid (TeA) and Radicinin (RAD), originated from Alternaria spp. Some of the toxins can be present in sorghum, ragi, wheat and tomatoes. Some research has shown that the toxins can be easily cross-contaminated between grain commodities, suggesting that manufacturing and storage of grain commodities is a critical practice.
- Citrinin.
- Citreoviridin.
- Cyclopiazonic acid.
- Cytochalasins.
- Ergot alkaloids / Ergopeptine alkaloids – Ergotamine.

- Fumonisins – Crop corn can be easily contaminated by the fungi Fusarium moniliforme, and its Fumonisin B1 will cause Leukoencephalomalacia (LEM) in horses, Pulmonary edema syndrome (PES) in pigs, liver cancer in rats and Esophageal cancer in humans. For human and animal health, both the FDA and the EC have regulated the content levels of toxins in food and animal feed.
- Fusaric acid.
- Fusarochromanone.
- Kojic acid.
- Lolitrem alkaloids.
- Moniliformin.
- 3-Nitropropionic acid.
- Nivalenol.
- Ochratoxins – In Australia, The Limit of Reporting (LOR) level for Ochratoxin A (OTA) analyses in 20th Australian Total Diet Survey was 1 μg/kg, whereas the EC restricts the content of OTA to 5 μg/kg in cereal commodities, 3 μg/kg in processed products and 10 μg/kg in dried vine fruits.
- Oosporeine.
- Patulin – Currently, this toxin has been advisably regulated on fruit products. The EC and the FDA have limited it to under 50 μg/kg for fruit juice and fruit nectar, while limits of 25 μg/kg for solid-contained fruit products and 10 μg/kg for baby foods were specified by the EC.
- Phomopsins.
- Sporidesmin A.
- Sterigmatocystin.
- Tremorgenic mycotoxins – Five of them have been reported to be associated with molds found in fermented meats. These are Fumitremorgen B, Paxilline, Penitrem A, Verrucosidin, and Verruculogen.
- Trichothecenes – sourced from Cephalosporium, Fusarium, Myrothecium, Stachybotrys and Trichoderma. The toxins are usually found in molded maize, wheat, corn, peanuts and rice, or animal feed of hay and straw. Four trichothecenes, T-2 toxin, HT-2 toxin, diacetoxyscirpenol (DAS) and deoxynivalenol (DON) have been most commonly encountered by humans and animals. The consequences of oral intake of, or dermal exposure to, the toxins will result in Alimentary toxic aleukia, neutropenia, aplastic anemia, thrombocytopenia and/or skin irritation. In 1993, the FDA issued a document for the content limits of DON in food and animal feed at an advisory level. In 2003, US published a patent that is very promising for farmers to produce a trichothecene-resistant crop.
- Zearalenone.

- Zearalenols.

Viruses

Viral infections make up perhaps one third of cases of food poisoning in developed countries. In the US, more than 50% of cases are viral and noroviruses are the most common foodborne illness, causing 57% of outbreaks in 2004. Foodborne viral infection are usually of intermediate (1–3 days) incubation period, causing illnesses which are self-limited in otherwise healthy individuals; they are similar to the bacterial forms described above.

- Enterovirus.
- Hepatitis A is distinguished from other viral causes by its prolonged (2–6 week) incubation period and its ability to spread beyond the stomach and intestines into the liver. It often results in jaundice, or yellowing of the skin, but rarely leads to chronic liver dysfunction. The virus has been found to cause infection due to the consumption of fresh-cut produce which has fecal contamination.
- Hepatitis E.
- Norovirus.
- Rotavirus.

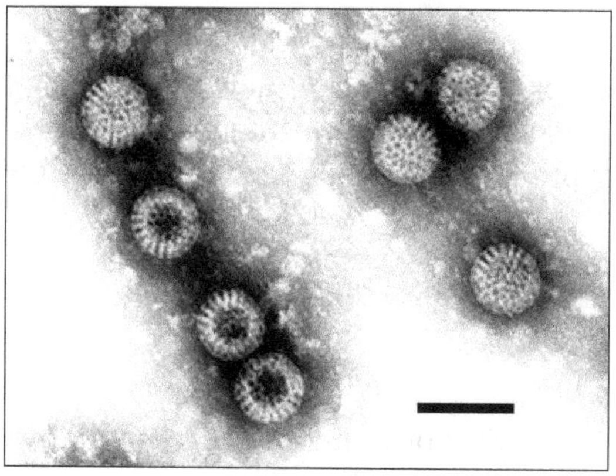

Rotavirus

Parasites

Most foodborne parasites are zoonoses.

- Platyhelminthes:
    - *Diphyllobothrium* sp.
    - *Nanophyetus* sp.
    - *Taenia saginata.*
    - *Taenia solium.*
    - *Fasciola hepatica.*

- Nematode:
  - o *Anisakis* sp.
  - o *Ascaris lumbricoides*.
  - o *Eustrongylides* sp.
  - o *Trichinella spiralis*.
  - o *Trichuris trichiura*.
- Protozoa:
  - o *Acanthamoeba* and other free-living amoebae·
  - o *Cryptosporidium parvum*.
  - o *Cyclospora cayetanensis*.
  - o *Entamoeba histolytica*.
  - o *Giardia lamblia*.

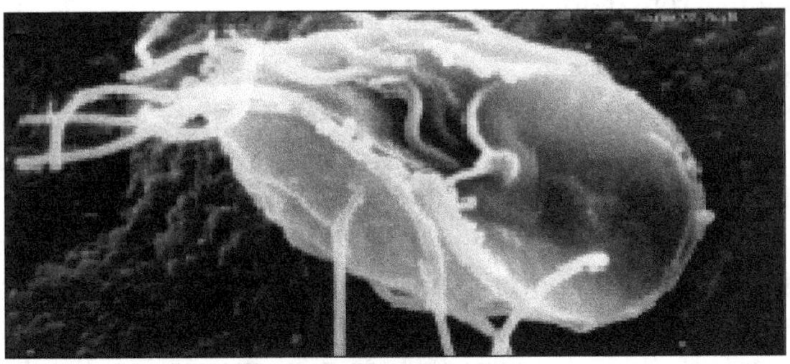

Giardia lamblia

  - o *Sarcocystis hominis*.
  - o *Sarcocystis suihominis*.
  - o *Toxoplasma gondii*.

Natural Toxins

Several foods can naturally contain toxins, many of which are not produced by bacteria. Plants in particular may be toxic; animals which are naturally poisonous to eat are rare. In evolutionary terms, animals can escape being eaten by fleeing; plants can use only passive defenses such as poisons and distasteful substances, for example capsaicin in chili peppers and pungent sulfur compounds in garlic and onions. Most animal poisons are not synthesised by the animal, but acquired by eating poisonous plants to which the animal is immune, or by bacterial action.

- Alkaloids.
- Ciguatera poisoning.
- Grayanotoxin (honey intoxication).

- Mushroom toxins.
- Phytohaemagglutinin (red kidney bean poisoning; destroyed by boiling).
- Pyrrolizidine alkaloids.
- Shellfish toxin, including paralytic shellfish poisoning, diarrhetic shellfish poisoning, neurotoxic shellfish poisoning, amnesic shellfish poisoning and ciguatera fish poisoning.
- Scombrotoxin.
- Tetrodotoxin (fugu fish poisoning).

Some plants contain substances which are toxic in large doses, but have therapeutic properties in appropriate dosages.

- Foxglove contains cardiac glycosides.
- Poisonous hemlock (conium) has medicinal uses.

Other Pathogenic Agents

- Prions, resulting in Creutzfeldt–Jakob disease.

Ptomaine Poisoning

In 1883, the Italian professor Salmi, of Bologna, introduced the generic name *ptomaine* (from Greek *ptōma*, "fall, fallen body, corpse") for alkaloids found in decaying animal and vegetable matter, especially (as reflected in their names) putrescine and cadaverine. The 1892 *Merck's Bulletin* stated, "We name such products of bacterial origin ptomaines; and the special alkaloid produced by the comma bacillus is variously named Cadaverine, Putrescine, etc." While *The Lancet* stated, "The chemical ferments produced in the system, the...ptomaines which may exercise so disastrous an influence." It is now known that the "disastrous...influence" is due to the direct action of bacteria and only slightly to the alkaloids. Thus, the use of the phrase "ptomaine poisoning" is now obsolete.

Tainted potato salad sickening hundreds at a Communist political convention in Massillon, Ohio, and aboard a Washington DC cruise boat in separate incidents during a single week in 1932 drew national attention to the dangers of so-called "ptomaine poisoning" in the pages of the American news weekly, *Time*. Another newspaper article from 1944 told of more than 150 persons being hospitalized in Chicago with ptomaine poisoning apparently from rice pudding served by a chain of restaurants.

Mechanism

Incubation Period

The delay between consumption of a contaminated food and appearance of the first symptoms of illness is called the incubation period. This ranges from hours to days (and rarely months or even years, such as in the case of listeriosis or bovine spongiform encephalopathy), depending on the agent, and on how much was consumed. If symptoms occur within one to six hours after eating the food, it suggests that it is caused by a bacterial toxin or a chemical rather than live bacteria.

The long incubation period of many foodborne illnesses tends to cause sufferers to attribute their symptoms to gastroenteritis.

During the incubation period, microbes pass through the stomach into the intestine, attach to the cells lining the intestinal walls, and begin to multiply there. Some types of microbes stay in the intestine, some produce a toxin that is absorbed into the bloodstream, and some can directly invade the deeper body tissues. The symptoms produced depend on the type of microbe.

Infectious Dose

The infectious dose is the amount of agent that must be consumed to give rise to symptoms of foodborne illness, and varies according to the agent and the consumer's age and overall health. In the case of *Salmonella* a relatively large inoculum of 1 million to 1 billion organisms is necessary to produce symptoms in healthy human volunteers, as *Salmonellae* are very sensitive to acid. An unusually high stomach pH level (low acidity) greatly reduces the number of bacteria required to cause symptoms by a factor of between 10 and 100.

Epidemiology

United States

In the United States, using FoodNet data from 2000–2007, the CDCP estimated there were 47.8 million foodborne illnesses per year (16,000 cases for 100,000 inhabitants) with 9.4 million of these caused by 31 known identified pathogens.

- 127,839 were hospitalized (43 per 100,000 inhabitants).
- 3,037 people died (1.0 per 100,000 inhabitants).

Causes of foodborne illness in U. S.			
	Cause	**Annual cases**	**Rate (per 100,000 inhabitants)**
1.	Norovirus	5,461,731 cases	X
2.	Salmonella	1,027,561 cases	X
3.	Clostridium perfringens	965,958 cases	X
4.	Campylobacter	845,024 cases	X

Causes of death by foodborne illness in U. S.			
	Cause	**Annual deaths**	**Rate (per 100,000 inhabitants)**
1.	Salmonella	378 cases	0.126
2.	Toxoplasma gondii	327 cases	0.109
3.	Listeria	255 cases	0.085
4.	Norovirus	149 cases	0.050

France

This data pertains to reported medical cases of 23 specific pathogens, as opposed to total population estimates of all food-borne illness for the United States.

In France, for 750,000 cases (1210 per 100,000 inhabitants):

- 70,000 people consulted in the emergency department of an hospital (113 per 100,000 inhabitants.).
- 113,000 people were hospitalized (182 per 100,000 inhabitants).
- 460 people died (0.75 per 100,000 inhabitants).

Causes of foodborne illness in France			
	Cause	**Annual hospitalizations**	**Rate (per 100,000 inhabitants)**
1.	Salmonella	~8,000 cases	13
2.	Campylobacter	~3,000 cases	4.8
3.	Parasites incl. Toxoplasma	~500 cases ~400 cases	0.8 0.65
4.	Listeria	~300 cases	0.5
5.	Hepatitis A	~60 cases	0.1

Causes of death by foodborne illness in France			
	Cause	**Annual**	**Rate (per 100,000 inhabitants)**
1.	Salmonella	~300 cases	0.5
2.	Listeria	~80 cases	0.13
3.	Parasites	~37 cases	0.06 (95% due to toxoplasma)
4.	Campylobacter	~15 cases	0.02
5.	Hepatitis A	~2 cases	0.003

Australia

A study by the Australian National University, published in November 2014, found in 2010 that there were an estimated 4.1 million cases of foodborne gastroenteritis acquired in Australia on average each year, along with 5,140 cases of non-gastrointestinal illness. The study was funded by the Australian Department of Health, Food Standards Australia New Zealand and the NSW Food Authority.

The main causes were Norovirus, pathogenic Escherichia coli, Campylobacter spp. and non-typhoidal Salmonella spp., although the causes of approximately 80% of illnesses were unknown. Approximately 25% (90% CrI: 13%–42%) of the 15.9 million episodes of gastroenteritis that occur in Australia were estimated to be transmitted by contaminated food. This equates to an average of approximately one episode of foodborne gastroenteritis every five years per person. Data on the number of hospitalisations and deaths represent the occurrence of serious foodborne illness. Including gastroenteritis, non-gastroenteritis and sequelae, there were an estimated annual 31,920 (90% CrI: 29,500–35,500) hospitalisations due to foodborne illness and 86 (90% CrI: 70–105) deaths due to foodborne illness circa 2010. This study concludes that these rates are similar to recent estimates in the US and Canada.

A main aim of this study was to compare if foodborne illness incidence had increased over time. In this study, similar methods of assessment were applied to data from circa 2000, which showed that the rate of foodborne gastroenteritis had not changed significantly over time. Two key estimates were the total number of gastroenteritis episodes each year, and the proportion considered foodborne. In circa 2010, it was estimated that 25% of all episodes of gastroenteritis were foodborne. By applying this proportion of episodes due to food to the incidence of gastroenteritis circa 2000, there were an estimated 4.3 million (90% CrI: 2.2–7.3 million) episodes of foodborne gastroenteritis circa 2000, although credible intervals overlap with 2010. Taking into account changes in population size, applying these equivalent methods suggests a 17% decrease in the rate of foodborne gastroenteritis between 2000 and 2010, with considerable overlap of the 90% credible intervals.

This study replaces a previous estimate of 5.4 million cases of food-borne illness in Australia every year, causing:

- 18,000 hospitalizations.
- 120 deaths (0.5 deaths per 100,000 inhabitants)
- 2.1 million lost days off work.
- 1.2 million doctor consultations.
- 300,000 prescriptions for antibiotics.

Most foodborne disease outbreaks in Australia have been linked to raw or minimally cooked eggs or poultry. The Australian Food Safety Information Council estimates that one third of cases of food poisoning occur in the home.

Comparison Between Countries

Country	Annual deaths per 100,000 inhabitants	Annual hospitalization per 100,000 inhabitants
USA	1.0	43
France	0.75	182
Australia	0.5	82

Outbreaks

The vast majority of reported cases of foodborne illness occur as individual or sporadic cases. The origin of most sporadic cases is undetermined. In the United States, where people eat outside the home frequently, 58% of cases originate from commercial food facilities (2004 FoodNet data). An outbreak is defined as occurring when two or more people experience similar illness after consuming food from a common source.

Often, a combination of events contributes to an outbreak, for example, food might be left at room temperature for many hours, allowing bacteria to multiply which is compounded by inadequate cooking which results in a failure to kill the dangerously elevated bacterial levels.

Outbreaks are usually identified when those affected know each other. However, more and more, outbreaks are identified by public health staff from unexpected increases in laboratory results for certain strains of bacteria. Outbreak detection and investigation in the United States is primarily handled by local health jurisdictions and is inconsistent from district to district. It is estimated that 1–2% of outbreaks are detected.

Society and Culture

United Kingdom

In postwar Aberdeen (1964) a large-scale (>400 cases) outbreak of typhoid occurred, caused by contaminated corned beef which had been imported from Argentina. The corned beef was placed in cans and because the cooling plant had failed, cold river water from the Plate estuary was used to cool the cans. One of the cans had a defect and the meat inside was contaminated. This meat was then sliced using a meat slicer in a shop in Aberdeen, and a lack of cleaning the machinery led to spreading the contamination to other meats cut in the slicer. These meats were then eaten by the people of Aberdeen who then became ill.

Serious outbreaks of foodborne illness since the 1970s prompted key changes in UK food safety law. These included the death of 19 patients in the Stanley Royd Hospital outbreak and the bovine spongiform encephalopathy (BSE, mad cow disease) outbreak identified in the 1980s. The death of 17 people in the 1996 Wishaw outbreak of *E. coli* O157 was a precursor to the establishment of the Food Standards Agency which, according to Tony Blair in the 1998 white paper *A Force for Change* Cm 3830, "would be powerful, open and dedicated to the interests of consumers".

In May 2015, for the second year running, England's Food Standards Agency devoted its annual Food Safety Week to – "The Chicken Challenge". The focus was on the handling of raw chicken in the home and in catering facilities in a drive to reduce the worryingly high levels of food poisoning from the *campylobacter* bacterium. Anne Hardy argues that widespread public education of food hygiene can be useful, particularly through media (T.V cookery programmes) and advertisement. She points to the examples set by Scandinavian societies.

United States

In 2001, the Center for Science in the Public Interest petitioned the United States Department of Agriculture to require meat packers to remove spinal cords before processing cattle carcasses for human consumption, a measure designed to lessen the risk of infection by variant Creutzfeldt-Jakob disease. The petition was supported by the American Public Health Association, the Consumer Federation of America, the Government Accountability Project, the National Consumers League, and Safe Tables Our Priority. This was opposed by the National Cattlemen's Beef Association, the National Renderers Association, the National Meat Association, the Pork Producers Council, sheep raisers, milk producers, the Turkey Federation, and eight other organizations from the animal-derived food industry.

None of the US Department of Health and Human Services targets regarding incidence of foodborne infections were reached in 2007.

Organizations

The World Health Organization Department of Food Safety and Zoonoses (FOS) provides scientific advice for organizations and the public on issues concerning the safety of food. Its mission is to lower the burden of foodborne disease, thereby strengthening the health security and sustainable development of Member States. Foodborne and waterborne diarrhoeal diseases kill an estimated 2.2 million people annually, most of whom are children. WHO works closely with the Food and Agriculture Organization of the United Nations (FAO) to address food safety issues along the entire food production chain—from production to consumption—using new methods of risk analysis. These methods provide efficient, science-based tools to improve food safety, thereby benefiting both public health and economic development.

International Food Safety Authorities Network (INFOSAN)

The International Food Safety Authorities Network (INFOSAN) is a joint program of the WHO and FAO. INFOSAN has been connecting national authorities from around the globe since 2004, with the goal of preventing the international spread of contaminated food and foodborne disease and strengthening food safety systems globally. This is done by:

- Promoting the rapid exchange of information during food safety events.
- Sharing information on important food safety issues of global interest.
- Promoting partnership and collaboration between countries.
- Helping countries strengthen their capacity to manage food safety risks.

Membership to INFOSAN is voluntary, but is restricted to representatives from national and regional government authorities and requires an official letter of designation. INFOSAN seeks to reflect the multidisciplinary nature of food safety and promote intersectoral collaboration by requesting the designation of Focal Points in each of the respective national authorities with a stake in food safety, and a single Emergency Contact Point in the national authority with the responsibility for coordinating national food safety emergencies; countries choosing to be members of INFOSAN are committed to sharing information between their respective food safety authorities and other INFOSAN members. The operational definition of a food safety authority includes those authorities involved in: food policy; risk assessment; food control and management; food inspection services; foodborne disease surveillance and response; laboratory services for monitoring and surveillance of foods and foodborne diseases; and food safety information, education and communication across the farm-to-table continuum.

Regulatory Steps

Food may be contaminated during all stages of food production and retailing. In order to prevent viral contamination, regulatory authorities in Europe have enacted several measures:

- European Commission Regulation (EC) No 2073/2005 of November 15, 2005.
- European Committee for Standardization (CEN): Standard method for the detection of norovirus and hepatitis A virus in food products (shellfish, fruits and vegetables, surfaces and bottled water).
- CODEX Committee on Food Hygiene (CCFH): Guideline for the application of general principles of food hygiene for the control of viruses in food.

Food Allergy

A food allergy is an abnormal immune response to food. The signs and symptoms may range from mild to severe. They may include itchiness, swelling of the tongue, vomiting, diarrhea, hives, trouble breathing, or low blood pressure. This typically occurs within minutes to several hours of exposure. When the symptoms are severe, it is known as anaphylaxis. Food intolerance and food poisoning are separate conditions.

Common foods involved include cow's milk, peanuts, eggs, shellfish, tree nuts, wheat, rice, and fruit. The common allergies in a region vary depending on the country. Risk factors include a family history of allergies, vitamin D deficiency, obesity, and high levels of cleanliness. Allergies occur when immunoglobulin E (IgE), part of the body's immune system, binds to food molecules. It is usually a protein in the food that is the problem. This triggers the release of inflammatory chemicals such as histamine. Diagnosis is usually based on a medical history, elimination diet, skin prick test, blood tests for food-specific IgE antibodies, or oral food challenge.

Early exposure to potential allergens may be protective. Management primarily involves avoiding the food in question and having a plan if exposure occurs. This plan may include giving adrenaline (epinephrine) and wearing medical alert jewelry. The benefits of allergen immunotherapy for food allergies is unclear and thus not recommended as of 2015. Some types of food allergies among children resolve with age including that to milk, eggs, and soy; while others such as to nuts and shellfish typically do not.

In the developed world about 4% to 8% of people have at least one food allergy. They are more common in children than adults and appear to be increasing in frequency. Male children appear to be more commonly affected than females. Some allergies more commonly develop early in life while others typically develop in later life. In developed countries, a large proportion of people believe they have food allergies when they actually do not have them.

Signs and Symptoms

Food allergies usually have a fast onset (from seconds to one hour) and may include:

- Rash.
- Hives.
- Itching of mouth, lips, tongue, throat, eyes, skin, or other areas.
- Swelling (angioedema) of lips, tongue, eyelids, or the whole face.
- Difficulty swallowing.
- Runny or congested nose.
- Hoarse voice.
- Wheezing and/or shortness of breath.
- Diarrhea, abdominal pain and/or stomach cramps.

- Lightheadedness.
- Fainting.
- Nausea.
- Vomiting.

In some cases, however, onset of symptoms may be delayed for hours.

Symptoms of allergies vary from person to person. The amount of food needed to trigger a reaction also varies from person to person.

Serious danger regarding allergies can begin when the respiratory tract or blood circulation is affected. The former can be indicated through wheezing and cyanosis. Poor blood circulation leads to a weak pulse, pale skin and fainting.

A severe case of an allergic reaction, caused by symptoms affecting the respiratory tract and blood circulation, is called anaphylaxis. When symptoms are related to a drop in blood pressure, the person is said to be in anaphylactic shock. Anaphylaxis occurs when IgE Antibodies are involved, and areas of the body that are not in direct contact with the food become affected and show symptoms. This occurs because no nutrients are circulated throughout the body, causing the widening of blood vessels. This vasodilation causes blood pressure to decrease, which leads to the loss of consciousness. Those with asthma or an allergy to peanuts, tree nuts, or seafood are at greater risk for anaphylaxis.

Cause

One of the most common food allergies is a sensitivity to peanuts, a member of the bean family. Peanut allergies may be severe, but children with peanut allergies sometimes outgrow them. Tree nuts, including cashews, Brazil nuts, hazelnuts, macadamia nuts, pecans, pistachios, pine nuts, coconuts, and walnuts, are also common allergens. Sufferers may be sensitive to one particular tree nut or to many different tree nuts. Also seeds, including sesame seeds and poppy seeds, contain oils where protein is present, which may elicit an allergic reaction.

Egg allergies affect about one in fifty children but are frequently outgrown by children when they reach age five. Typically the sensitivity is to proteins in the white, rather than the yolk.

Milk, from cows, goats or sheep, is another common food allergen, and many sufferers are also unable to tolerate dairy products such as cheese. A small portion of children with a milk allergy, roughly ten percent, will have a reaction to beef. Beef contains a small amount of protein that is present in cow's milk.

Other foods containing allergenic proteins include soy, wheat, fish, shellfish, fruits, vegetables, maize, spices, synthetic and natural colors, and chemical additives.

Although sensitivity levels vary by country, the most common food allergies are allergies to milk, eggs, peanuts, tree nuts, seafood, shellfish, soy and wheat. These are often referred to as "the big eight." Allergies to seeds — especially sesame — seem to be increasing in many countries. An example an allergy more common to a particular region is that to rice in East Asia where rice forms a large part of the diet.

Balsam of Peru, which is in various foods, is in the "top five" allergens most commonly causing patch test reactions in people referred to dermatology clinics.

Sensitization

An Institute of Medicine report says that food proteins contained in vaccines, such as gelatin, milk or egg can cause sensitization (development of allergy) in vaccine recipients, to those food items.

Atopy

Food allergies develop more easily in people with the atopic syndrome, a very common combination of diseases: allergic rhinitis and conjunctivitis, eczema and asthma. The syndrome has a strong inherited component; a family history of allergic diseases can be indicative of the atopic syndrome.

Cross Reactivity

Some children who are allergic to cow's milk protein also show a cross sensitivity to soy-based products. There are infant formulas in which the milk and soy proteins are degraded so when taken by an infant, their immune system does not recognize the allergen and they can safely consume the product. Hypoallergenic infant formulas can be based on hydrolyzed proteins, which are proteins partially predigested in a less antigenic form. Other formulas, based on free amino acids, are the least antigenic and provide complete nutrition support in severe forms of milk allergy.

People with latex allergy often also develop allergies to bananas, kiwifruit, avocados, and some other foods.

Pathophysiology

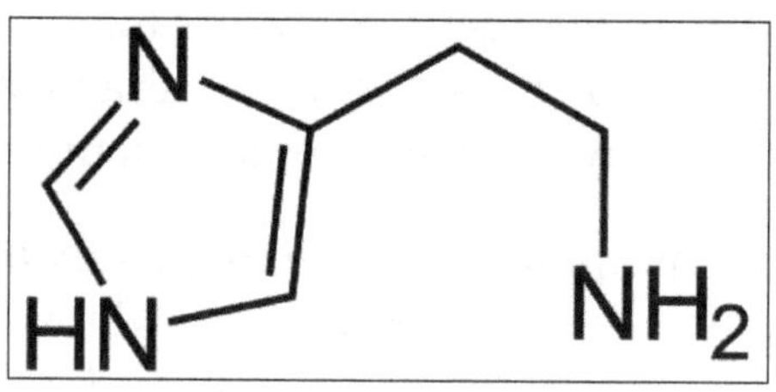

A histamine, the structure shown, is what causes a person to feel itchy during an allergic reaction. A common medication to stop this is an antihistamine, which fights the histamines in the person's system.

A food allergy occurs when the human immune system reacts abnormally to certain foods.

Conditions caused by food allergies are classified into 3 groups according to the mechanism of the allergic response:

1. IgE-mediated (classic) – the most common type, occurs shortly after eating and may involve anaphylaxis.
2. Non-IgE mediated – characterized by an immune response not involving immunoglobulin E; may occur some hours after eating, complicating diagnosis.
3. IgE and/or non-IgE-mediated – a hybrid of the above two types.

Allergic reactions are hyperactive responses of the immune system to generally innocuous substances. When immune cells encounter the allergenic protein, IgE antibodies are produced; this is similar to the immune system's reaction to foreign pathogens. The IgE antibodies identify the allergenic proteins as harmful and initiate the allergic reaction. The harmful proteins are those that do not break down due to the strong bonds of the protein. IgE antibodies bind to a receptor on the surface of the protein, creating a tag, just as a virus or parasite becomes tagged. It is not entirely clear why some proteins do not denature and subsequently trigger allergic reactions and hypersensitivity while others do.

Hypersensitivities are categorized according to the parts of the immune system that are attacked and the amount of time it takes for the response to occur. There are four types of Hypersensitivity reaction: Type 1, Immediate IgE-mediated, Type 2, Cytotoxic, Type 3, Immune complex-mediated, and Type 4, Delayed cell-mediated. The pathophysiology of allergic responses can be divided into two phases. The first is an acute response that occurs immediately after exposure to an allergen. This phase can either subside or progress into a "late phase reaction" which can substantially prolong the symptoms of a response, and result in tissue damage.

Many food allergies are caused by hypersensitivities to particular proteins in different foods. Proteins have unique properties that allow them to become allergens, such as stabilizing forces in the tertiary and quaternary structure which prevent degradation during digestion. Many theoretically allergenic proteins cannot survive the destructive environment of the digestive tract and thus don't trigger hypersensitive reactions.

Acute Response

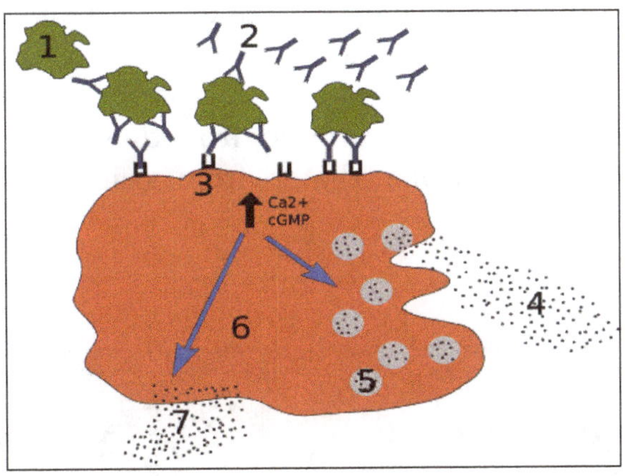

Degranulation process in allergy.
1 — antigen.
2 — IgE antibody.
3 — FcεRI receptor.
4 — preformed mediators (histamine, proteases, chemokines, heparin).
5 — granules.
6 — mast cell.
7 — newly formed mediators (prostaglandins, leukotrienes, thromboxanes, PAF).

In the early stages of allergy, a type I hypersensitivity reaction against an allergen, encountered for the first time, causes a response in a type of immune cell called a T_H2 lymphocyte, which belongs to a subset of T cells that produce a cytokine called interleukin-4 (IL-4). These T_H2 cells interact with other lymphocytes called B cells, whose role is the production of antibodies. Coupled with signals provided by IL-4, this interaction stimulates the B cell to begin production of a large amount of a particular type of antibody known as IgE. Secreted IgE circulates in the blood and binds to an IgE-specific receptor (a kind of Fc receptor called FcεRI) on the surface of other kinds of immune cells called mast cells and basophils, which are both involved in the acute inflammatory response. The IgE-coated cells, at this stage are sensitized to the allergen.

If later exposure to the same allergen occurs, the allergen can bind to the IgE molecules held on the surface of the mast cells or basophils. Cross-linking of the IgE and Fc receptors occurs when more than one IgE-receptor complex interacts with the same allergenic molecule, and activates the sensitized cell. Activated mast cells and basophils undergo a process called degranulation, during which they release histamine and other inflammatory chemical mediators (cytokines, interleukins, leukotrienes, and prostaglandins) from their granules into the surrounding tissue causing several systemic effects, such as vasodilation, mucous secretion, nerve stimulation and smooth muscle contraction. This results in rhinorrhea, itchiness, dyspnea, and anaphylaxis. Depending on the individual, the allergen, and the mode of introduction, the symptoms can be system-wide (classical anaphylaxis), or localized to particular body systems; asthma is localized to the respiratory system and eczema is localized to the dermis.

Late-phase Response

After the chemical mediators of the acute response subside, late phase responses can often occur. This is due to the migration of other leukocytes such as neutrophils, lymphocytes, eosinophils, and macrophages to the initial site. The reaction is usually seen 2–24 hours after the original reaction. Cytokines from mast cells may also play a role in the persistence of long-term effects. Late phase responses seen in asthma are slightly different from those seen in other allergic responses, although they are still caused by release of mediators from eosinophils, and are still dependent on activity of T_H2 cells.

Diagnosis

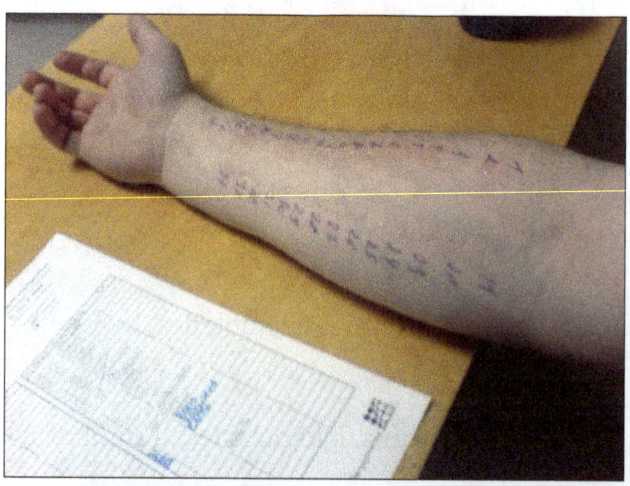

Skin testing on the arm is a common way for detecting an allergy, however, it is not as effective as other tests.

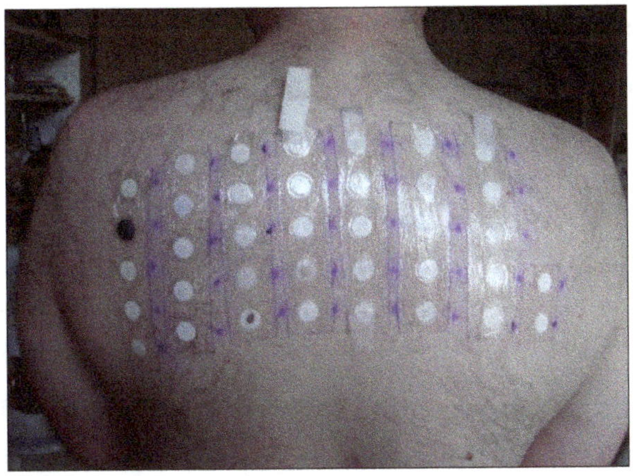

Patch test

Diagnosis is usually based on a medical history, elimination diet, skin prick test, blood tests for food-specific IgE antibodies, or oral food challenge.

- For skin prick tests, a tiny board with protruding needles is used. The allergens are placed either on the board or directly on the skin. The board is then placed on the skin, in order to puncture the skin and for the allergens to enter the body. If a hive appears, the person will be considered positive for the allergy. This test only works for IgE antibodies. Allergic reactions caused by other antibodies cannot be detected through skin prick tests.

Skin prick testing is easy to do and results are available in minutes. Different allergists may use different devices for skin prick testing. Some use a "bifurcated needle", which looks like a fork with 2 prongs. Others use a "multi-test", which may look like a small board with several pins sticking out of it. In these tests, a tiny amount of the suspected allergen is put onto the skin or into a testing device, and the device is placed on the skin to prick, or break through, the top layer of skin. This puts a small amount of the allergen under the skin. A hive will form at any spot where the person is allergic. This test generally yields a positive or negative result. It is good for quickly learning if a person is allergic to a particular food or not, because it detects allergic antibodies known as IgE. Skin tests cannot predict if a reaction would occur or what kind of reaction might occur if a person ingests that particular allergen. They can however confirm an allergy in light of a patient's history of reactions to a particular food. Non-IgE mediated allergies cannot be detected by this method.

- Patch testing is a method used to determine if a specific substance causes allergic inflammation of the skin. It tests for delayed food reactions.
- Blood testing is another way to test for allergies; however, it poses the same disadvantage and only detects IgE allergens and does not work for every possible allergen. RAST, RadioAllergoSorbent Test, is used to detect IgE antibodies present to a certain allergen. The score taken from the RAST test is compared to predictive values, taken from a specific type of RAST test. If the score is higher than the predictive values, there is a great chance the allergy is present in the person. One advantage of this test is that it can test many allergens at one time.

Blood tests are another useful diagnostic tool for evaluating IgE-mediated food allergies. For example, the RAST (RadioAllergoSorbent Test) detects the presence of IgE antibodies to a particular allergen. A CAP-RAST test is a specific type of RAST test with greater specificity: it can show the amount of IgE present to each allergen. Researchers have been able to determine "predictive values" for certain foods. These predictive values can be compared to the RAST blood test results. If a persons RAST score is higher than the predictive value for that food, then there is over a 95% chance the person will have an allergic reaction (limited to rash and anaphylaxis reactions) if they ingest that food. Currently, predictive values are available for the following foods: milk, egg, peanut, fish, soy, and wheat. Blood tests allow for hundreds of allergens to be screened from a single sample, and cover food allergies as well as inhalants. However, non-IgE mediated allergies cannot be detected by this method. Other widely promoted tests such as the *antigen leukocyte cellular antibody test* (ALCAT) and the *Food Allergy Profile* are considered unproven methods, the use of which is not advised.

- Food challenges test for allergens other than those caused by IgE allergens. The allergen is given to the person in the form of a pill, so the person can ingest the allergen directly. The person is watched for signs and symptoms. The problem with food challenges is that they must be performed in the hospital under careful watch, due to the possibility of anaphylaxis.

Food challenges, especially double-blind placebo-controlled food challenges (DBPCFC), are the gold standard for diagnosis of food allergies, including most non-IgE mediated reactions. Blind food challenges involve packaging the suspected allergen into a capsule, giving it to the patient, and observing the patient for signs or symptoms of an allergic reaction.

The best method for diagnosing food allergy is to be assessed by an allergist. The allergist will review the patient's history and the symptoms or reactions that have been noted after food ingestion. If the allergist feels the symptoms or reactions are consistent with food allergy, he/she will perform allergy tests.

- Additional diagnostic tools for evaluation of eosinophilic or non-IgE mediated reactions include endoscopy, colonoscopy, and biopsy.

Differential Diagnosis

Important differential diagnoses are:

- Lactose intolerance; this generally develops later in life but can present in young patients in severe cases. This is due to an enzyme deficiency (lactase) and not allergy. It occurs in many non-Western people.

- Celiac disease; this is an autoimmune disorder triggered by gluten proteins such as gliadin (present in wheat, rye and barley). It is a non-IgE mediated food allergy by definition.

- Irritable bowel syndrome (IBS).

- C1 esterase inhibitor deficiency (hereditary angioedema); this rare disease generally causes attacks of angioedema, but can present solely with abdominal pain and occasional diarrhea.

Prevention

Breastfeeding for more than 4 months may prevent atopic dermatitis, cow milk allergy, and wheezing in early childhood. Early exposure to potential allergens may be protective.

In order to avoid an allergic reaction, a strict diet can be followed. It is difficult to determine the amount of allergenic food required to elicit a reaction, so complete avoidance should be attempted. In some cases, hypersensitive reactions can be triggered by exposures to allergens through skin contact, inhalation, kissing, participation in sports, blood transfusions, cosmetics, and alcohol.

Treatment

The mainstay of treatment for food allergy is total avoidance of the foods that have been identified as allergens. An allergen can enter the body by consuming a food containing the allergen, and can also be ingested by touching any surfaces that may have come into contact with the allergen, then touching the eyes or nose. For people who are extremely sensitive, avoidance includes avoiding touching or inhaling the problematic food.

If the food is accidentally ingested and a systemic reaction (anaphylaxis) occurs, then epinephrine should be used. It is possible that a second dose of epinephrine may be required for severe reactions. The person should then be transported to the emergency room, where additional treatment can be given. Other treatments include antihistamines and steroids.

Epinephrine

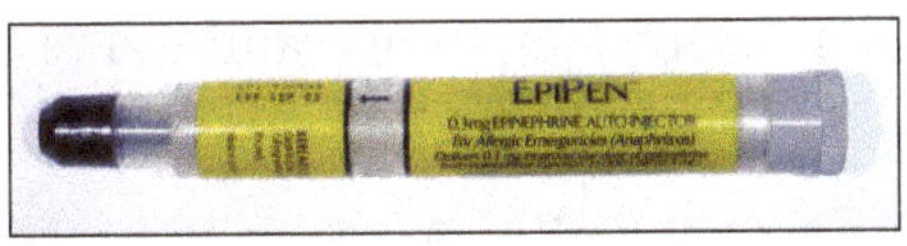

Epinephrine auto-injectors are portable single-dose epinephrine-dispensing devices used to treat anaphylaxis.

Epinephrine is another name for the hormone adrenaline, which is produced naturally in the body. An epinephrine injection is the first-line treatment for severe allergic reactions (known as anaphylaxis). If administered in a timely manner, epinephrine can reverse the effects of anaphylaxis.

Epinephrine relieves airway swelling and obstruction and improves blood circulation; blood vessels are tightened and heart rate is increased, improving circulation to bodily organs. Epinephrine is available by prescription in an auto-injector which is used in the treatment of anaphylaxis.

Antihistamines

Antihistamines can alleviate some of the milder symptoms of an allergic reaction, but do not treat all symptoms of anaphylaxis. Antihistamines block the action of histamine, which causes blood vessels to dilate and become leaky to plasma proteins. Histamine also causes itchiness by acting on sensory nerve terminals. The most common antihistamine given for food allergies is diphenhydramine.

Steroids

Glucocorticoid steroids are used to calm down the immune system cells that are attacked by the chemicals released during an allergic reaction. This form of treatment in the form of a nasal spray should not be used to treat anaphylaxis, for it only relieves symptoms in the area in which the steroid is in contact. Another reason steroids should not be used to treat anaphylaxis is due to the long amount of time it takes to reduce inflammation and start to work. Steroids can also be taken orally or through injection. By taking a steroid in these manners, every part of the body can be reached and treated, but a long time is usually needed for these to take effect.

Desensitization

The benefits of allergen immunotherapy for food allergies is unclear and thus not recommended as of 2015. A number of desensitization techniques are; however, being studied.

Epidemiology

The most common food allergens include peanuts, milk, eggs, tree nuts, fish, shellfish, soy, and wheat — these foods account for about 90% of all allergic reactions. The most common food allergies in adults are shellfish, peanuts, tree nuts, fish, and egg. The most common food allergies in children are milk, eggs, peanuts, and tree nuts.

Six to eight percent of children under the age of three have food allergies and nearly four percent of adults have food allergies.

For reasons that are not entirely understood, the diagnosis of food allergies has apparently become more common in Western nations in recent times. In the United States, food allergy affects as many as 5% of infants less than three years of age and 3% to 4% of adults. There is a similar prevalence in Canada.

Seventy-five percent of children who have allergies to milk protein are able to tolerate baked-in milk products, i.e., muffins, cookies, cake and hydrolysed formulas.

About 50% of children with allergies to milk, egg, soy, peanuts, tree nuts and wheat will outgrow their allergy by the age of 6. Those that are still allergic by the age of 12 or so have less than an 8% chance of outgrowing the allergy.

Peanut and tree nut allergies are less likely to be outgrown, although evidence now shows that about 20% of those with peanut allergies and 9% of those with tree nut allergies will outgrow them.

In Central Europe, celery allergy is more common. In Japan, allergy to buckwheat flour, used for soba noodles, is more common.

Meat allergy is extremely rare in the general population, but a geographic cluster of people allergic to meat has been observed in Sydney, Australia. There appears to be a possible association between localised reaction to tick bite and the development of meat allergy.

Fruit allergies exist, such as to: citrus fruits, apples, peaches, pears, jackfruit, strawberries, etc.. This is suspected to be associated with ragweed pollen allergy but could be due to other factors. Reactions can be severe.

Corn allergy may also be prevalent in many populations, although it may be difficult to recognize in areas such as the United States and Canada where corn derivatives are common in the food supply.

Protein allergies or intolerance of seeds, nuts, meat, and milk are especially common among children.

United States

In the United States, it is estimated that up to twelve million people have food allergies.

Food allergies cause roughly 30,000 emergency room visits and 100 to 200 deaths per year and the prevalence is rising. Food allergy affects as many as 5% of infants less than three years of age and 3% to 4% of adults.

For example, an estimated 8% of children have at least one food allergy, a projected total of 5.9 million children across the population, according to a 2011 study published in the journal *Pediatrics*, "The Prevalence, Severity, and Distribution of Childhood Food Allergy in the United States."

Although sensitivity levels vary by country, the most common food allergies are allergies to milk, eggs, peanuts, tree nuts, seafood, shellfish, soy and wheat. These are often referred to as "the big eight." They account for over 90% of the food allergies in the United States.

Society and Culture

In response to the risk that certain foods pose to those with food allergies, some countries have responded by instituting labeling laws that require food products to clearly inform consumers if their products contain major allergens or by-products of major allergens. Some countries also require companies to warn customers when food has been prepared around certain allergens that have been known to cause severe reactions.

From 13 December 2014, new legislation – the EU Food Information for Consumers Regulation 1169/2011 – requires food businesses to provide allergy information on food sold unpackaged, in for example catering outlets, deli counters, bakeries and sandwich bars.

There is much discretion on what food allergies are, and many members of society still believe that a food allergy is simply a dislike of a certain food, and do not realize that it is an actual medical condition.

Law

An example of a list of allergens in a food item.

Under the Food Allergen Labeling and Consumer Protection Act of 2004 (Public Law 108-282), companies are required to disclose on the label whether the product contains a major food allergen in clear, plain language. The allergens have to clearly be called out in the ingredient statement. Most companies list allergens in a statement separate from the ingredient statement.

In 2009, Governor Deval Patrick signed into Massachusetts law an Act Relative to Food Allergy Awareness in Restaurants. The allergy awareness act requires food protection managers to view a video about food allergens, a poster identifying the 8 most common food allergens and information about identifying and responding to food allergies posted for food service staff, and customers must be notified of their obligation to inform staff about any food allergies.

On 4 January 2011, President Barack Obama signed into federal law the Food Safety and Modernization Act of 2010 (S510/HR2751, 111th Congress). Section 112 of this Act establishes voluntary food allergy and anaphylaxis management guidelines for public kindergartens, elementary and secondary schools.

Research

Areas of research include anti-IgE antibody (omalizumab) and specific oral tolerance induction (SOTI), which have shown some promise for treatment of certain food allergies.

Gastroenteritis

Gastroenteritis, also known as infectious diarrhea, is inflammation of the gastrointestinal tract that involves the stomach and small intestine. Signs and symptoms include some combination of diarrhea, vomiting, and abdominal pain. Fever, lack of energy, and dehydration may also occur. This typically lasts less than two weeks. It is unrelated to influenza though it has been called the stomach flu.

Gastroenteritis can be due to infections by viruses, bacteria, parasites, and fungus. The most common cause is viruses. In children rotavirus is the most common cause of severe disease. In adults, norovirus and *Campylobacter* are common. Transmission may occur due to eating improperly prepared foods, drinking contaminated water, or through close contact with an individual who is infected. Testing to confirm the diagnosis is typically not needed.

Prevention includes hand washing with soap, drinking clean water, proper disposal of human waste, and breastfeeding babies instead of using formula. The rotavirus vaccine is recommended in children. Treatment involves getting enough fluids. For mild or moderate cases, this can typically be achieved by drinking oral rehydration solution (a combination of water, salts, and sugar). In those who are breast fed, continued breastfeeding is recommended. For more severe cases, intravenous fluids may be needed. Fluids may also be given by a nasogastric tube. Zinc supplementation is recommended in children. Antibiotics are generally not needed.

It is estimated that three to five billion cases of gastroenteritis resulting in 1.4 million deaths occur globally each year. Children and those in the developing world are most commonly affected. As of 2011, in those below age five, there were about 1.7 billion cases resulting in 0.7 million deaths. In the developing world children less than two years of age frequently get six or more infections a year. It is less common in adults, partly due to the development of immunity.

Signs and Symptoms

Gastroenteritis typically involves both diarrhea and vomiting, or less commonly, presents with only one or the other. Abdominal cramping may also be present. Signs and symptoms usually begin 12–72 hours after contracting the infectious agent. If due to a viral agent, the condition usually resolves within one week. Some viral causes may also be associated with fever, fatigue, headache, and muscle pain. If the stool is bloody, the cause is less likely to be viral and more likely to be bacterial. Some bacterial infections may be associated with severe abdominal pain and may persist for several weeks.

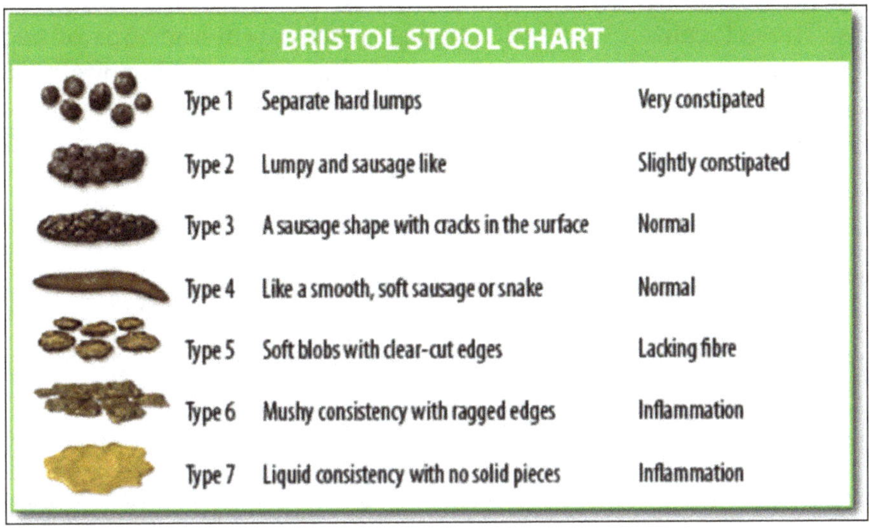

Bristol stool chart.

Children infected with rotavirus usually make a full recovery within three to eight days. However, in poor countries treatment for severe infections is often out of reach and persistent diarrhea is common. Dehydration is a common complication of diarrhea, and a child with a significant degree of dehydration may have a prolonged capillary refill, poor skin turgor, and abnormal breathing. Repeat infections are typically seen in areas with poor sanitation, and malnutrition, stunted growth, and long-term cognitive delays can result.

Reactive arthritis occurs in 1% of people following infections with *Campylobacter* species, and Guillain-Barre syndrome occurs in 0.1%. Hemolytic uremic syndrome (HUS) may occur due to infection with Shiga toxin-producing *Escherichia coli* or *Shigella* species, causing low platelet counts, poor kidney function, and low red blood cell count (due to their breakdown). Children are more predisposed to getting HUS than adults. Some viral infections may produce benign infantile seizures.

Cause

Viruses (particularly rotavirus) and the bacteria *Escherichia coli* and *Campylobacter* species are the primary causes of gastroenteritis. There are, however, many other infectious agents that can cause this syndrome. Non-infectious causes are seen on occasion, but they are less likely than a viral or bacterial cause. Risk of infection is higher in children due to their lack of immunity and relatively poor hygiene.

Viral

Rotavirus, norovirus, adenovirus, and astrovirus are known to cause viral gastroenteritis. Rotavirus is the most common cause of gastroenteritis in children, and produces similar rates in both the developed and developing world. Viruses cause about 70% of episodes of infectious diarrhea in the pediatric age group. Rotavirus is a less common cause in adults due to acquired immunity. Norovirus is the cause in about 18% of all cases.

Norovirus is the leading cause of gastroenteritis among adults in America, causing greater than 90% of outbreaks. These localized epidemics typically occur when groups of people spend time in close physical proximity to each other, such as on cruise ships, in hospitals, or in restaurants. People may remain infectious even after their diarrhea has ended. Norovirus is the cause of about 10% of cases in children.

Bacterial

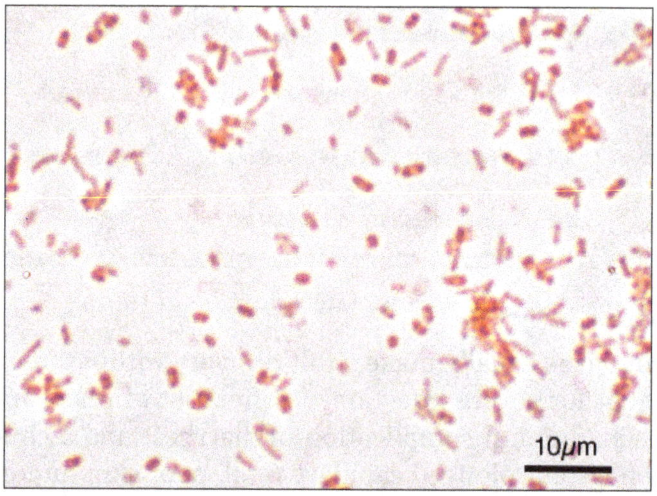

Salmonella enterica serovar Typhimurium (ATCC 14028) as seen with a microscope at 1000 fold magnification and following Gram staining.

In the developed world *Campylobacter jejuni* is the primary cause of bacterial gastroenteritis, with half of these cases associated with exposure to poultry. In children, bacteria are the cause in about 15% of cases, with the most common types being *Escherichia coli, Salmonella, Shigella,* and *Campylobacter* species. If food becomes contaminated with bacteria and remains at room temperature for a period of several hours, the bacteria multiply and increase the risk of infection in those who consume the food. Some foods commonly associated with illness include raw or undercooked meat, poultry, seafood, and eggs; raw sprouts; unpasteurized milk and soft cheeses; and fruit and vegetable juices. In the developing world, especially sub-Saharan Africa and Asia, cholera is a common cause of gastroenteritis. This infection is usually transmitted by contaminated water or food.

Toxigenic *Clostridium difficile* is an important cause of diarrhea that occurs more often in the elderly. Infants can carry these bacteria without developing symptoms. It is a common cause of diarrhea in those who are hospitalized and is frequently associated with antibiotic use. *Staphylococcus aureus* infectious diarrhea may also occur in those who have used antibiotics. Acute "traveler's diarrhea" is usually a type of bacterial gastroenteritis, while the persistent form is usually parasitic. Acid-suppressing medication appears to increase the risk of significant infection after exposure to a number of organisms, including *Clostridium difficile, Salmonella,* and *Campylobacter* species. The risk is greater in those taking proton pump inhibitors than with H2 antagonists.

Parasitic

A number of protozoa can cause gastroenteritis – most commonly *Giardia lamblia* – but *Entamoeba histolytica*, *Cryptosporidium* spp., and other species have also been implicated. As a group, these agents comprise about 10% of cases in children. *Giardia* occurs more commonly in the developing world, but this etiologic agent causes this type of illness to some degree nearly everywhere. It occurs more commonly in persons who have traveled to areas with high prevalence, children who attend day care, men who have sex with men, and following disasters.

Transmission

Transmission may occur via consumption of contaminated water, or when people share personal objects. In places with wet and dry seasons, water quality typically worsens during the wet season, and this correlates with the time of outbreaks. In areas of the world with four seasons, infections are more common in the winter. Bottle-feeding of babies with improperly sanitized bottles is a significant cause on a global scale. Transmission rates are also related to poor hygiene, especially among children, in crowded households, and in those with pre-existing poor nutritional status. After developing tolerance, adults may carry certain organisms without exhibiting signs or symptoms, and thus act as natural reservoirs of contagion. While some agents (such as *Shigella*) only occur in primates, others may occur in a wide variety of animals (such as *Giardia*).

Non-infectious

There are a number of non-infectious causes of inflammation of the gastrointestinal tract. Some of the more common include medications (like NSAIDs), certain foods such as lactose (in those who are intolerant), and gluten (in those with celiac disease). Crohn's disease is also a non-infectious source of (often severe) gastroenteritis. Disease secondary to toxins may also occur. Some food-related conditions associated with nausea, vomiting, and diarrhea include: ciguatera poisoning due to consumption of contaminated predatory fish, scombroid associated with the consumption of certain types of spoiled fish, tetrodotoxin poisoning from the consumption of puffer fish among others, and botulism typically due to improperly preserved food.

In the United States, rates of emergency department use for noninfectious gastroenteritis dropped 30% from 2006 until 2011. Of the twenty most common conditions seen in the emergency department, rates of noninfectious gastroenteritis had the largest decrease in visits in that time period.

Pathophysiology

Gastroenteritis is defined as vomiting or diarrhea due to infection of the small or large bowel. The changes in the small bowel are typically noninflammatory, while the ones in the large bowel are inflammatory. The number of pathogens required to cause an infection varies from as few as one (for *Cryptosporidium*) to as many as 10^8 (for *Vibrio cholerae*).

Diagnosis

Gastroenteritis is typically diagnosed clinically, based on a person's signs and symptoms. Determining the exact cause is usually not needed as it does not alter management of the condition.

However, stool cultures should be performed in those with blood in the stool, those who might have been exposed to food poisoning, and those who have recently traveled to the developing world. Diagnostic testing may also be done for surveillance. As hypoglycemia occurs in approximately 10% of infants and young children, measuring serum glucose in this population is recommended. Electrolytes and kidney function should also be checked when there is a concern about severe dehydration.

Dehydration

A determination of whether or not the person has dehydration is an important part of the assessment, with dehydration typically divided into mild (3–5%), moderate (6–9%), and severe (≥10%) cases. In children, the most accurate signs of moderate or severe dehydration are a prolonged capillary refill, poor skin turgor, and abnormal breathing. Other useful findings (when used in combination) include sunken eyes, decreased activity, a lack of tears, and a dry mouth. A normal urinary output and oral fluid intake is reassuring. Laboratory testing is of little clinical benefit in determining the degree of dehydration. Thus the use of urine testing or ultrasounds is generally not needed.

Differential Diagnosis

Other potential causes of signs and symptoms that mimic those seen in gastroenteritis that need to be ruled out include appendicitis, volvulus, inflammatory bowel disease, urinary tract infections, and diabetes mellitus. Pancreatic insufficiency, short bowel syndrome, Whipple's disease, coeliac disease, and laxative abuse should also be considered. The differential diagnosis can be complicated somewhat if the person exhibits *only* vomiting or diarrhea (rather than both).

Appendicitis may present with vomiting, abdominal pain, and a small amount of diarrhea in up to 33% of cases. This is in contrast to the large amount of diarrhea that is typical of gastroenteritis. Infections of the lungs or urinary tract in children may also cause vomiting or diarrhea. Classical diabetic ketoacidosis (DKA) presents with abdominal pain, nausea, and vomiting, but without diarrhea. One study found that 17% of children with DKA were initially diagnosed as having gastroenteritis.

Prevention

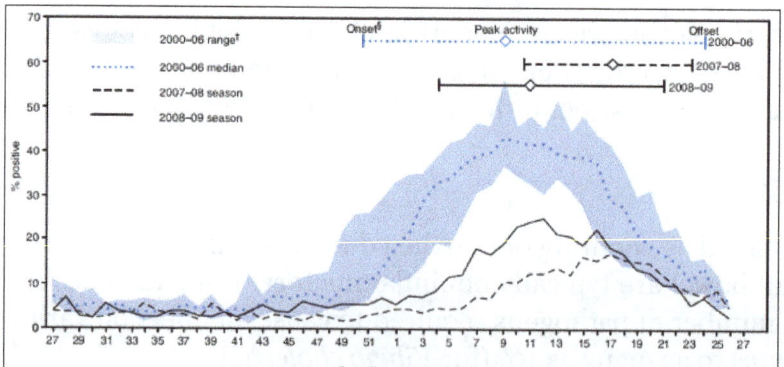

Percentage of rotavirus tests with positive results, by surveillance week, United States, July 2000 – June 2009.

Lifestyle

A supply of easily accessible uncontaminated water and good sanitation practices are important for reducing rates of infection and clinically significant gastroenteritis. Personal measures (such as hand washing with soap) have been found to decrease rates of gastroenteritis in both the developing and developed world by as much as 30%. Alcohol-based gels may also be effective. Breastfeeding is important, especially in places with poor hygiene, as is improvement of hygiene generally. Breast milk reduces both the frequency of infections and their duration. Avoiding contaminated food or drink should also be effective.

Vaccination

Due to both its effectiveness and safety, in 2009 the World Health Organization recommended that the rotavirus vaccine be offered to all children globally. Two commercial rotavirus vaccines exist and several more are in development. In Africa and Asia these vaccines reduced severe disease among infants and countries that have put in place national immunization programs have seen a decline in the rates and severity of disease. This vaccine may also prevent illness in non-vaccinated children by reducing the number of circulating infections. Since 2000, the implementation of a rotavirus vaccination program in the United States has substantially decreased the number of cases of diarrhea by as much as 80 percent. The first dose of vaccine should be given to infants between 6 and 15 weeks of age. The oral cholera vaccine has been found to be 50–60% effective over 2 years.

Management

Gastroenteritis is usually an acute and self-limiting disease that does not require medication. The preferred treatment in those with mild to moderate dehydration is oral rehydration therapy (ORT). Metoclopramide and/or ondansetron, however, may be helpful in some children, and butylscopolamine is useful in treating abdominal pain.

Rehydration

The primary treatment of gastroenteritis in both children and adults is rehydration. This is preferably achieved by drinking rehydration solution, although intravenous delivery may be required if there is a decreased level of consciousness or if dehydration is severe. Drinking replacement therapy products made with complex carbohydrates (i.e. those made from wheat or rice) may be superior to those based on simple sugars. Drinks especially high in simple sugars, such as soft drinks and fruit juices, are not recommended in children under 5 years of age as they may *increase* diarrhea. Plain water may be used if more specific ORT preparations are unavailable or the person is not willing to drink them. A nasogastric tube can be used in young children to administer fluids if warranted. In those who require intravenous fluids, one to four hours' worth is often sufficient.

Dietary

It is recommended that breast-fed infants continue to be nursed in the usual fashion, and that formula-fed infants continue their formula immediately after rehydration with ORT. Lactose-free or lactose-reduced formulas usually are not necessary. Children should continue their usual diet during episodes of diarrhea with the exception that foods high in simple sugars should be avoided. The BRAT diet (bananas, rice, applesauce, toast and tea) is no longer recommended, as it contains insufficient nutrients and has no benefit over normal feeding. Some probiotics have been shown to be beneficial in reducing both the duration of illness and the frequency of stools. They may also be useful in preventing and treating antibiotic associated diarrhea. Fermented milk products (such as yogurt) are similarly beneficial. Zinc supplementation appears to be effective in both treating and preventing diarrhea among children in the developing world.

Antiemetics

Antiemetic medications may be helpful for treating vomiting in children. Ondansetron has some utility, with a single dose being associated with less need for intravenous fluids, fewer hospitalizations, and decreased vomiting. Metoclopramide might also be helpful. However, the use of ondansetron might possibly be linked to an increased rate of return to hospital in children. The intravenous preparation of ondansetron may be given orally if clinical judgment warrants. Dimenhydrinate, while reducing vomiting, does not appear to have a significant clinical benefit.

Antibiotics

Antibiotics are not usually used for gastroenteritis, although they are sometimes recommended if symptoms are particularly severe or if a susceptible bacterial cause is isolated or suspected. If antibiotics are to be employed, a macrolide (such as azithromycin) is preferred over a fluoroquinolone due to higher rates of resistance to the latter. Pseudomembranous colitis, usually caused by antibiotic use, is managed by discontinuing the causative agent and treating it with either metronidazole or vancomycin. Bacteria and protozoans that are amenable to treatment include *Shigella Salmonella typhi*, and *Giardia* species. In those with *Giardia* species or *Entamoeba histolytica*, tinidazole treatment is recommended and superior to metronidazole. The World Health Organization (WHO) recommends the use of antibiotics in young children who have both bloody diarrhea and fever.

Antimotility Agents

Antimotility medication has a theoretical risk of causing complications, and although clinical experience has shown this to be unlikely, these drugs are discouraged in people with bloody diarrhea or diarrhea that is complicated by fever. Loperamide, an opioid analogue, is commonly used for the symptomatic treatment of diarrhea. Loperamide is not recommended in children, however, as it may cross the immature blood–brain barrier and cause toxicity. Bismuth subsalicylate, an insoluble complex of trivalent bismuth and salicylate, can be used in mild to moderate cases, but salicylate toxicity is theoretically possible.

Epidemiology

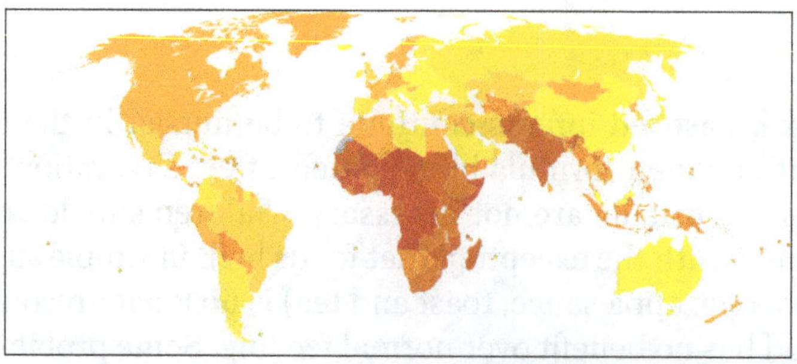

Deaths due to diarrhoeal diseases per million persons in 2012.

- 0-2
- 3-10
- 11-18
- 19-30
- 31-46
- 47-80
- 81-221
- 222-450
- 451-606
- 607-1799

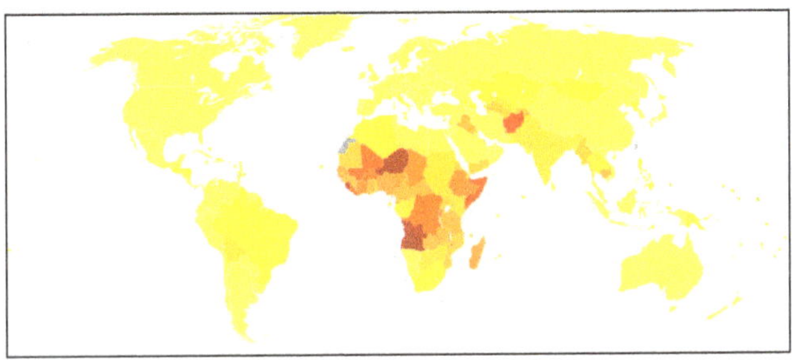

Disability-adjusted life year for diarrhea per 100,000 inhabitants in 2004.

no data	3000–3500
≤less 500	3500–4000
500–1000	4000–4500
1000–1500	4500–5000
1500–2000	5000–6000
2000–2500	≥6000
2500–3000	

It is estimated that three to five billion cases of gastroenteritis resulting in 1.4 million deaths occur globally on an annual basis. Children and those in the developing world are most commonly affected. As of 2011, in those less than five, there were about 1.7 billion cases resulting in 0.7 million deaths, with most of these occurring in the world's poorest nations. More than 450,000 of these fatalities are due to rotavirus in children under 5 years of age. Cholera causes about three to five million cases of disease and kills approximately 100,000 people yearly. In the developing world, children less than two years of age frequently get six or more infections a year that result in significant gastroenteritis. It is less common in adults, partly due to the development of acquired immunity.

In 1980, gastroenteritis from all causes caused 4.6 million deaths in children, with the majority occurring in the developing world. Death rates were reduced significantly (to approximately 1.5 million deaths annually) by the year 2000, largely due to the introduction and widespread use of oral rehydration therapy. In the US, infections causing gastroenteritis are the second most common infection (after the common cold), and they result in between 200 and 375 million cases of acute diarrhea and approximately ten thousand deaths annually, with 150 to 300 of these deaths in children less than five years of age.

History

The first usage of "gastroenteritis" was in 1825. Before this time it was commonly known as typhoid fever or "cholera morbus", among others, or less specifically as "griping of the guts", "surfeit", "flux", "colic", "bowel complaint", or any one of a number of other archaic names for acute diarrhea. Cholera morbus is a historical term that was used to refer to gastroenteritis rather than specifically cholera.

Society and Culture

Gastroenteritis is associated with many colloquial names, including "Montezuma's revenge", "Delhi belly", "la turista", and "back door sprint", among others. It has played a role in many military campaigns and is believed to be the origin of the term "no guts no glory".

Gastroenteritis is the main reason for 3.7 million visits to physicians a year in the United States and 3 million visits in France. In the United States gastroenteritis as a whole is believed to result in costs of 23 billion USD per year with that due to rotavirus alone resulting in estimated costs of 1 billion USD a year.

Research

There are a number of vaccines against gastroenteritis in development. For example, vaccines against *Shigella* and enterotoxigenic *Escherichia coli* (ETEC), two of the leading bacterial causes of gastroenteritis worldwide.

Other Animals

Many of the same agents cause gastroenteritis in cats and dogs as in humans. The most common organisms are *Campylobacter*, *Clostridium difficile*, *Clostridium perfringens*, and *Salmonella*. A large number of toxic plants may also cause symptoms.

Some agents are more specific to a certain species. Transmissible gastroenteritis coronavirus (TGEV) occurs in pigs resulting in vomiting, diarrhea, and dehydration. It is believed to be introduced to pigs by wild birds and there is no specific treatment available. It is not transmissible to humans.

Diarrhea

Diarrhea, also spelled diarrhoea, is the condition of having at least three loose or liquid bowel movements each day. It often lasts for a few days and can result in dehydration due to fluid loss.

Signs of dehydration often begin with loss of the normal stretchiness of the skin and irritable behaviour. This can progress to decreased urination, loss of skin color, a fast heart rate, and a decrease in responsiveness as it becomes more severe. Loose but non-watery stools in babies who are breastfed, however, may be normal.

The most common cause is an infection of the intestines due to either a virus, bacteria, or parasite; a condition known as gastroenteritis. These infections are often acquired from food or water that has been contaminated by stool, or directly from another person who is infected. It may be divided into three types: short duration watery diarrhea, short duration bloody diarrhea, and if it lasts for more than two weeks, persistent diarrhea. The short duration watery diarrhea may be due to an infection by cholera, although this is rare in the developed world. If blood is present it is also known as dysentery. A number of non-infectious causes may also result in diarrhea, including hyperthyroidism, lactose intolerance, inflammatory bowel disease, a number of medications, and irritable bowel syndrome. In most cases, stool cultures are not required to confirm the exact cause.

Prevention of infectious diarrhea is by improved sanitation, clean drinking water, and hand washing with soap. Breastfeeding for at least six months is also recommended as is vaccination against rotavirus. Oral rehydration solution (ORS), which is clean water with modest amounts of salts and sugar, is the treatment of choice. Zinc tablets are also recommended. These treatments have been estimated to have saved 50 million children in the past 25 years. When people have diarrhea it is recommended that they continue to eat healthy food and babies continue to be breastfed. If commercial ORS are not available, homemade solutions may be used. In those with severe dehydration, intravenous fluids may be required. Most cases; however, can be managed well with fluids by mouth. Antibiotics, while rarely used, may be recommended in a few cases such as those who have bloody diarrhea and a high fever, those with severe diarrhea following travelling, and those who grow specific bacteria or parasites in their stool. Loperamide may help decrease the number of bowel movements but is not recommended in those with severe disease.

About 1.7 to 5 billion cases of diarrhea occur per year. It is most common in developing countries, where young children get diarrhea on average three times a year. Total deaths from diarrhea are estimated at 1.26 million in 2013 – down from 2.58 million in 1990. In 2012, it was the second most common cause of deaths in children younger than five (0.76 million or 11%). Frequent episodes of diarrhea are also a common cause of malnutrition and the most common cause in those younger than five years of age. Other long term problems that can result include stunted growth and poor intellectual development.

Definition

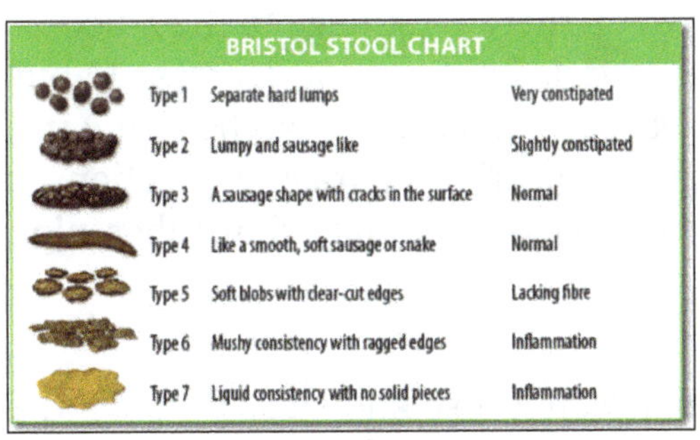

Bristol stool chart.

Diarrhea is defined by the World Health Organization as having three or more loose or liquid stools per day, or as having more stools than is normal for that person.

Acute diarrhea is defined as an abnormally frequent discharge of semisolid or fluid fecal matter from the bowel, lasting less than 14 days, by World Gastroenterology Organization.

Secretory

Secretory diarrhea means that there is an increase in the active secretion, or there is an inhibition of absorption. There is little to no structural damage. The most common cause of this type of diarrhea is a cholera toxin that stimulates the secretion of anions, especially chloride ions. Therefore, to maintain a charge balance in the gastrointestinal tract, sodium is carried with it, along with water. In this type of diarrhea intestinal fluid secretion is isotonic with plasma even during fasting. It continues even when there is no oral food intake.

Osmotic

Osmotic diarrhea occurs when too much water is drawn into the bowels. If a person drinks solutions with excessive sugar or excessive salt, these can draw water from the body into the bowel and cause osmotic diarrhea. Osmotic diarrhea can also be the result of maldigestion (e.g., pancreatic disease or Coeliac disease), in which the nutrients are left in the lumen to pull in water. Or it can be caused by osmotic laxatives (which work to alleviate constipation by drawing water into the bowels). In healthy individuals, too much magnesium or vitamin C or undigested lactose can produce osmotic diarrhea and distention of the bowel. A person who has lactose intolerance can have difficulty absorbing lactose after an extraordinarily high intake of dairy products. In persons who have fructose malabsorption, excess fructose intake can also cause diarrhea. High-fructose foods that also have a high glucose content are more absorbable and less likely to cause diarrhea. Sugar alcohols such as sorbitol (often found in sugar-free foods) are difficult for the body to absorb and, in large amounts, may lead to osmotic diarrhea. In most of these cases, osmotic diarrhea stops when the offending agent (e.g. milk, sorbitol) is stopped.

Exudative

Exudative diarrhea occurs with the presence of blood and pus in the stool. This occurs with inflammatory bowel diseases, such as Crohn's disease or ulcerative colitis, and other severe infections such as *E. coli* or other forms of food poisoning.

Inflammatory

Inflammatory diarrhea occurs when there is damage to the mucosal lining or brush border, which leads to a passive loss of protein-rich fluids and a decreased ability to absorb these lost fluids. Features of all three of the other types of diarrhea can be found in this type of diarrhea. It can be caused by bacterial infections, viral infections, parasitic infections, or autoimmune problems such as inflammatory bowel diseases. It can also be caused by tuberculosis, colon cancer, and enteritis.

Dysentery

If there is blood visible in the stools, it is also known as dysentery. The blood is a trace of an invasion of bowel tissue. Dysentery is a symptom of, among others, *Shigella*, *Entamoeba histolytica*, and *Salmonella*.

Health Effects

Diarrheal disease may have a negative impact on both physical fitness and mental development. "Early childhood malnutrition resulting from any cause reduces physical fitness and work productivity in adults," and diarrhea is a primary cause of childhood malnutrition. Further, evidence suggests that diarrheal disease has significant impacts on mental development and health; it has been shown that, even when controlling for helminth infection and early breastfeeding, children who had experienced severe diarrhea had significantly lower scores on a series of tests of intelligence.

Differential Diagnosis

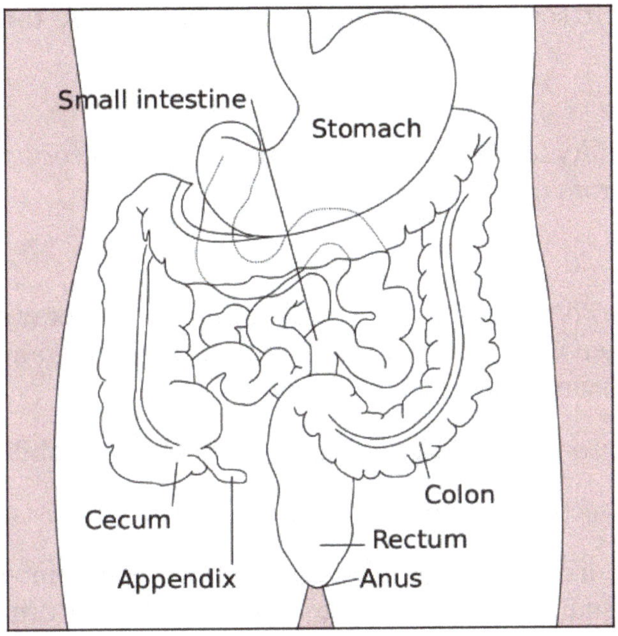

Diagram of the human gastrointestinal tract.

Acute diarrhea is most commonly due to viral gastroenteritis with rotavirus, which accounts for 40% of cases in children under five. (p. 17) In travelers however bacterial infections predominate. Various toxins such as mushroom poisoning and drugs can also cause acute diarrhea.

Chronic diarrhea can be the part of the presentations of a number of chronic medical conditions affecting the intestine. Common causes include ulcerative colitis, Crohn's disease, microscopic colitis, celiac disease, irritable bowel syndrome and bile acid malabsorption.

Infections

There are many causes of infectious diarrhea, which include viruses, bacteria and parasites. Infectious diarrhea is frequently referred to as gastroenteritis. Norovirus is the most common cause of viral diarrhea in adults, but rotavirus is the most common cause in children under five years old. Adenovirus types 40 and 41, and astroviruses cause a significant number of infections.

Campylobacter spp. are a common cause of bacterial diarrhea, but infections by *Salmonella* spp., *Shigella* spp. and some strains of *Escherichia coli* are also a frequent cause.

In the elderly, particularly those who have been treated with antibiotics for unrelated infections, a toxin produced by *Clostridium difficile* often causes severe diarrhea.

Parasites, particularly protozoa (e.g., *Cryptosporidium* spp., *Giardia* spp., *Entamoeba histolytica*, *Blastocystis* spp., *Cyclospora cayetanensis*), are frequently the cause of diarrhea that involves chronic infection. The broad-spectrum antiparasitic agent nitazoxanide has shown efficacy against many diarrhea-causing parasites.

Other infectious agents, such as parasites or bacterial toxins, may exacerbate symptoms. In sanitary living conditions where there is ample food and a supply of clean water, an otherwise healthy person usually recovers from viral infections in a few days. However, for ill or malnourished individuals, diarrhea can lead to severe dehydration and can become life-threatening.

Malabsorption

Malabsorption is the inability to absorb food fully, mostly from disorders in the small bowel, but also due to maldigestion from diseases of the pancreas.

Causes include:

- *Enzyme deficiencies or mucosal abnormality*, as in food allergy and food intolerance, e.g. celiac disease (gluten intolerance), lactose intolerance (intolerance to milk sugar, common in non-Europeans), and fructose malabsorption.
- *Pernicious anemia*, or impaired bowel function due to the inability to absorb vitamin B_{12}.
- *Loss of pancreatic secretions*, which may be due to cystic fibrosis or pancreatitis.
- *Structural defects*, like short bowel syndrome (surgically removed bowel) and radiation fibrosis, such as usually follows cancer treatment and other drugs, including agents used in chemotherapy.
- *Certain drugs*, like orlistat, which inhibits the absorption of fat.

Inflammatory Bowel Disease

The two overlapping types here are of unknown origin:

- Ulcerative colitis is marked by chronic bloody diarrhea and inflammation mostly affects the distal colon near the rectum.
- Crohn's disease typically affects fairly well demarcated segments of bowel in the colon and often affects the end of the small bowel.

Irritable Bowel Syndrome

Another possible cause of diarrhea is irritable bowel syndrome (IBS), which usually presents with abdominal discomfort relieved by defecation and unusual stool (diarrhea or constipation) for at least 3 days a week over the previous 3 months. Symptoms of diarrhea-predominant IBS can be managed through a combination of dietary changes, soluble fiber supplements, and/or medications such as loperamide or codeine. About 30% of patients with diarrhea-predominant IBS have bile acid malabsorption diagnosed with an abnormal SeHCAT test.

Other Diseases

Diarrhea can be caused by other diseases and conditions, namely:

- Chronic ethanol ingestion.
- Ischemic bowel disease: This usually affects older people and can be due to blocked arteries.
- Microscopic colitis, a type of inflammatory bowel disease where changes are only seen on histological examination of colonic biopsies.
- Bile salt malabsorption (primary bile acid diarrhea) where excessive bile acids in the colon produce a secretory diarrhea.
- Hormone-secreting tumors: some hormones (e.g., serotonin) can cause diarrhea if excreted in excess (usually from a tumor).
- Chronic mild diarrhea in infants and toddlers may occur with no obvious cause and with no other ill effects; this condition is called toddler's diarrhea.
- Environmental enteropathy.
- Radiation enteropathy following treatment for pelvic and abdominal cancers.

Causes

Sanitation

Poverty often leads to unhygienic living conditions, as in this community in the Indian Himalayas. Such conditions promote contraction of diarrheal diseases, as a result of poor sanitation and hygiene.

Open defecation is a leading cause of infectious diarrhea leading to death.

Poverty is a good indicator of the rate of infectious diarrhea in a population. This association does not stem from poverty itself, but rather from the conditions under which impoverished people live. The absence of certain resources compromises the ability of the poor to defend themselves against infectious diarrhea. "Poverty is associated with poor housing, crowding, dirt floors, lack of access to clean water or to sanitary disposal of fecal waste (sanitation), cohabitation with domestic animals that may carry human pathogens, and a lack of refrigerated storage for food, all of which increase the frequency of diarrhea... Poverty also restricts the ability to provide age-appropriate, nutritionally balanced diets or to modify diets when diarrhea develops so as to mitigate and repair nutrient losses. The impact is exacerbated by the lack of adequate, available, and affordable medical care."

Water

One of the most common causes of infectious diarrhea, is a lack of clean water. Often, improper fecal disposal leads to contamination of groundwater. This can lead to widespread infection among a population, especially in the absence of water filtration or purification. Human feces contains a variety of potentially harmful human pathogens.

Nutrition

Proper nutrition is important for health and functioning, including the prevention of infectious diarrhea. It is especially important to young children who do not have a fully developed immune system. Zinc deficiency, a condition often found in children in developing countries can, even in mild cases, have a significant impact on the development and proper functioning of the human immune system. Indeed, this relationship between zinc deficiency and reduced immune functioning corresponds with an increased severity of infectious diarrhea. Children who have lowered levels of zinc have a greater number of instances of diarrhea, severe diarrhea, and diarrhea associated with fever. Similarly, vitamin A deficiency can cause an increase in the severity of diarrheal episodes. However, there is some discrepancy when it comes to the impact of vitamin A deficiency on the rate of disease. While some argue that a relationship does not exist between the rate of disease and vitamin A status, others suggest an increase in the rate associated with deficiency. Given that estimates suggest 127 million preschool children worldwide are vitamin A deficient, this population has the potential for increased risk of disease contraction.

Pathophysiology

Evolution

According to two researchers, Nesse and Williams, diarrhea may function as an evolved expulsion defense mechanism. As a result, if it is stopped, there might be a delay in recovery. They cite in support of this argument research published in 1973 that found that treating *Shigella* with the anti-diarrhea drug (Co-phenotrope, Lomotil) caused people to stay feverish twice as long as those not so treated. The researchers indeed themselves observed that: "Lomotil may be contraindicated in shigellosis. Diarrhea may represent a defense mechanism".

Diagnostic Approach

The following types of diarrhea may indicate further investigation is needed:

- In infants.

- Moderate or severe diarrhea in young children.
- Associated with blood.
- Continues for more than two days.
- Associated non-cramping abdominal pain, fever, weight loss, etc.
- In travelers.
- In food handlers, because of the potential to infect others.
- In institutions such as hospitals, child care centers, or geriatric and convalescent homes.

A severity score is used to aid diagnosis in children.

Prevention

Sanitation

Numerous studies have shown that improvements in drinking water and sanitation (WASH) lead to decreased risks of diarrhoea. Such improvements might include for example use of water filters, provision of high-quality piped water and sewer connections.

In institutions, communities, and households, interventions that promote hand washing with soap lead to significant reductions in the incidence of diarrhea. The same applies to preventing open defecation at a community-wide level and providing access to improved sanitation. This includes use of toilets and implementation of the entire sanitation chain connected to the toilets (collection, transport, disposal or reuse of human excreta).

Hand Washing

Basic sanitation techniques can have a profound effect on the transmission of diarrheal disease. The implementation of hand washing using soap and water, for example, has been experimentally shown to reduce the incidence of disease by approximately 42–48%. Hand washing in developing countries, however, is compromised by poverty as acknowledged by the CDC: "Handwashing is integral to disease prevention in all parts of the world; however, access to soap and water is limited in a number of less developed countries. This lack of access is one of many challenges to proper hygiene in less developed countries." Solutions to this barrier require the implementation of educational programs that encourage sanitary behaviours.

Water

Given that water contamination is a major means of transmitting diarrheal disease, efforts to provide clean water supply and improved sanitation have the potential to dramatically cut the rate of disease incidence. In fact, it has been proposed that we might expect an 88% reduction in child mortality resulting from diarrheal disease as a result of improved water sanitation and hygiene. Similarly, a meta-analysis of numerous studies on improving water supply and sanitation shows a 22–27% reduction in disease incidence, and a 21–30% reduction in mortality rate associated with diarrheal disease.

Chlorine treatment of water, for example, has been shown to reduce both the risk of diarrheal disease, and of contamination of stored water with diarrheal pathogens.

Vaccination

Immunization against the pathogens that cause diarrheal disease is a viable prevention strategy, however it does require targeting certain pathogens for vaccination. In the case of Rotavirus, which was responsible for around 6% of diarrheal episodes and 20% of diarrheal disease deaths in the children of developing countries, use of a Rotavirus vaccine in trials in 1985 yielded a slight (2-3%) decrease in total diarrheal disease incidence, while reducing overall mortality by 6-10%. Similarly, a Cholera vaccine showed a strong reduction in morbidity and mortality, though the overall impact of vaccination was minimal as Cholera is not one of the major causative pathogens of diarrheal disease. Since this time, more effective vaccines have been developed that have the potential to save many thousands of lives in developing nations, while reducing the overall cost of treatment, and the costs to society.

A rotavirus vaccine decrease the rates of diarrhea in a population. New vaccines against rotavirus, *Shigella*, Enterotoxigenic Escherichia coli (ETEC), and cholera are under development, as well as other causes of infectious diarrhea.

Nutrition

Dietary deficiencies in developing countries can be combated by promoting better eating practices. Supplementation with vitamin A and/or zinc. Zinc supplementation proved successful showing a significant decrease in the incidence of diarrheal disease compared to a control group. The majority of the literature suggests that vitamin A supplementation is advantageous in reducing disease incidence. Development of a supplementation strategy should take into consideration the fact that vitamin A supplementation was less effective in reducing diarrhea incidence when compared to vitamin A and zinc supplementation, and that the latter strategy was estimated to be significantly more cost effective.

Breastfeeding

Breastfeeding practices have been shown to have a dramatic effect on the incidence of diarrheal disease in poor populations. Studies across a number of developing nations have shown that those who receive exclusive breastfeeding during their first 6 months of life are better protected against infection with diarrheal diseases. Exclusive breastfeeding is currently recommended during, at least, the first six months of an infant's life by the WHO.

Others

Probiotics decrease the risk of diarrhea in those taking antibiotics.

Management

In many cases of diarrhea, replacing lost fluid and salts is the only treatment needed. This is usually by mouth – oral rehydration therapy – or, in severe cases, intravenously. Diet restrictions such as the BRAT diet are no longer recommended. Research does not support the limiting of milk to children as doing so has no effect on duration of diarrhea. To the contrary, WHO recommends that children with diarrhea continue to eat as sufficient nutrients are usually still absorbed to support continued growth and weight gain, and that continuing to eat also speeds up recovery of normal intestinal functioning. CDC recommends that children and adults with cholera also continue to eat.

Medications such as loperamide (Imodium) and bismuth subsalicylate may be beneficial; however they may be contraindicated in certain situations.

Fluids

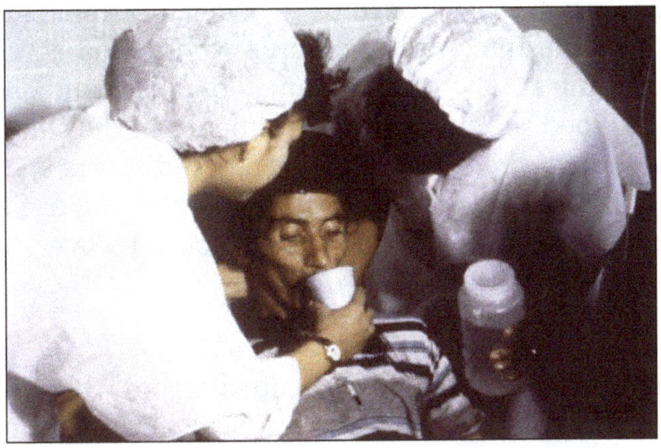

A person consuming oral rehydration solution.

Oral rehydration solution (ORS) (a slightly sweetened and salty water) can be used to prevent dehydration. Standard home solutions such as salted rice water, salted yogurt drinks, vegetable and chicken soups with salt can be given. Home solutions such as water in which cereal has been cooked, unsalted soup, green coconut water, weak tea (unsweetened), and unsweetened fresh fruit juices can have from half a teaspoon to full teaspoon of salt (from one-and-a-half to three grams) added per liter. Clean plain water can also be one of several fluids given. There are commercial solutions such as Pedialyte, and relief agencies such as UNICEF widely distribute packets of salts and sugar. A WHO publication for physicians recommends a homemade ORS consisting of one liter water with one teaspoon salt (3 grams) and two tablespoons sugar (18 grams) added (approximately the "taste of tears"). Rehydration Project recommends adding the same amount of sugar but only one-half a teaspoon of salt, stating that this more dilute approach is less risky with very little loss of effectiveness. Both agree that drinks with too much sugar or salt can make dehydration worse.

Appropriate amounts of supplemental zinc and potassium should be added if available. But the availability of these should not delay rehydration. As WHO points out, the most important thing is to begin preventing dehydration as early as possible. In another example of prompt ORS hopefully preventing dehydration, CDC recommends for the treatment of cholera continuing to give Oral Rehydration Solution during travel to medical treatment.

Vomiting often occurs during the first hour or two of treatment with ORS, especially if a child drinks the solution too quickly, but this seldom prevents successful rehydration since most of the fluid is still absorbed. WHO recommends that if a child vomits, to wait five or ten minutes and then start to give the solution again more slowly.

Drinks especially high in simple sugars, such as soft drinks and fruit juices, are not recommended in children under 5 years of age as they may *increase* dehydration. A too rich solution in the gut draws water from the rest of the body, just as if the person were to drink sea water. Plain water may be used if more specific and effective ORT preparations are unavailable or are not palatable. Additionally, a mix of both plain water and drinks perhaps too rich in sugar and salt can alternatively be given to the same person, with the goal of providing a medium amount of sodium overall. A nasogastric tube can be used in young children to administer fluids if warranted.

Eating

WHO recommends a child with diarrhea continue to be fed. Continued feeding speeds the recovery of normal intestinal function. In contrast, children whose food is restricted have diarrhea of longer duration and recover intestinal function more slowly. A child should also continue to be breastfed. The WHO states "Food should *never* be withheld and the child's usual foods should *not* be diluted. Breastfeeding should *always* be continued." And in the specific example of cholera, CDC also makes the same recommendation. In young children who are not breast-fed and live in the developed world, a lactose-free diet may be useful to speed recovery.

Medications

While antibiotics are beneficial in certain types of acute diarrhea, they are usually not used except in specific situations. There are concerns that antibiotics may increase the risk of hemolytic uremic syndrome in people infected with Escherichia coli O157:H7. In resource-poor countries, treatment with antibiotics may be beneficial. However, some bacteria are developing antibiotic resistance, particularly *Shigella*. Antibiotics can also cause diarrhea, and antibiotic-associated diarrhea is the most common adverse effect of treatment with general antibiotics.

While bismuth compounds (Pepto-Bismol) decreased the number of bowel movements in those with travelers' diarrhea, they do not decrease the length of illness. Anti-motility agents like loperamide are also effective at reducing the number of stools but not the duration of disease. These agents should only be used if bloody diarrhea is not present.

Bile acid sequestrants such as cholestyramine can be effective in chronic diarrhea due to bile acid malabsorption. Therapeutic trials of these drugs are indicated in chronic diarrhea if bile acid malabsorption cannot be diagnosed with a specific test, such as SeHCAT retention.

Alternative Therapies

Zinc supplementation benefits children with diarrhea in developing countries, but only in infants over six months old. This supports the World Health Organization guidelines for zinc, but not in the very young.

Probiotics reduce the duration of symptoms by one day and reduced the chances of symptoms lasting longer than four days by 60%. The probiotic lactobacillus can help prevent antibiotic-associated diarrhea in adults but possibly not children. For those with lactose intolerance, taking digestive enzymes containing lactase when consuming dairy products often improves symptoms.

Epidemiology

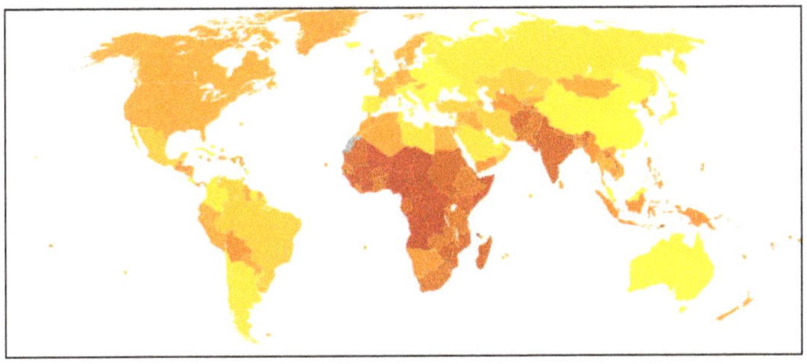

Deaths due to diarrhoeal diseases per million persons in 2012.

- 0-2
- 3-10
- 11-18
- 19-30
- 31-46
- 47-80
- 81-221
- 222-450
- 451-606
- 607-1799

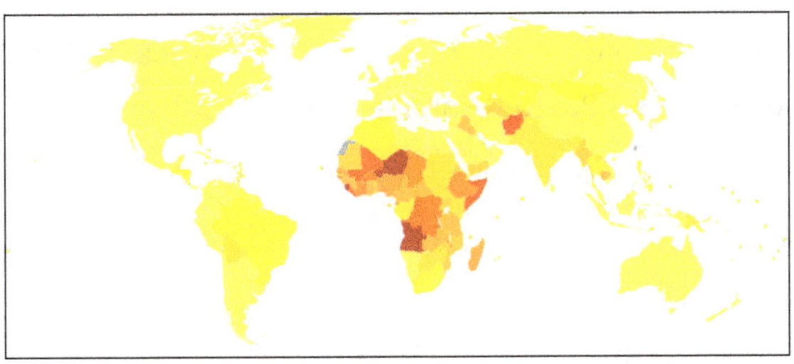

Disability-adjusted life year for diarrhea per 100,000 inhabitants in 2004.

- no data
- < 500
- 500–1000
- 1000–1500
- 1500–2000

- 2000–2500
- 2500–3000
- 3000–3500
- 3500–4000
- 4000–4500
- 4500–5000
- 5000–6000
- > 6000

Worldwide in 2004, approximately 2.5 billion cases of diarrhea occurred, which resulted in 1.5 million deaths among children under the age of five. Greater than half of these were in Africa and South Asia. This is down from a death rate of 4.5 million in 1980 for gastroenteritis. Diarrhea remains the second leading cause of infant mortality (16%) after pneumonia (17%) in this age group.

The majority of such cases occur in the developing world, with over half of the recorded cases of childhood diarrhea occurring in Africa and Asia, with 696 million and 1.2 billion cases, respectively, compared to only 480 million in the rest of the world.

Infectious diarrhea resulted in about 0.7 million deaths in children under five years old in 2011 and 250 million lost school days. In the Americas, diarrheal disease accounts for a total of 10% of deaths among children aged 1–59 months while in South East Asia, it accounts for 31.3% of deaths. It is estimated that around 21% of child mortalities in developing countries are due to diarrheal disease.

Etymology

The word diarrhea is from “through” and *rheo* “flow”. Diarrhea is the spelling in American English while diarrhoea is the spelling in Commonwealth English.

Nausea

Nausea is a sensation of unease and discomfort in the upper stomach with an involuntary urge to vomit. It may precede vomiting, but a person can have nausea without vomiting. When prolonged, it is a debilitating symptom.

Nausea is a non-specific symptom, which means that it has many possible causes. Some common causes of nausea are motion sickness, dizziness, migraine, fainting, gastroenteritis (stomach infection) or food poisoning. Nausea is a side effect of many medications including chemotherapy, nauseants or morning sickness in early pregnancy. Nausea may also be caused by anxiety, disgust and depression.

Medications taken to prevent and treat nausea are called antiemetics. The most commonly prescribed antiemetics in the US are promethazine, metoclopramide and ondansetron. The word nausea is from Latin *nausea*, "motion sickness", "feeling sick or queasy".

Causes

There are many causes of nausea. One organization listed 700 in 2009. Gastrointestinal infections (37%) and food poisoning are the two most common causes. Side effects from medications (3%) and pregnancy are also relatively frequent. In 10% of people the cause remains unknown.

Food Poisoning

Food poisoning usually causes an abrupt onset of nausea and vomiting one to six hours after ingestion of contaminated food and lasts for one to two days. It is due to toxins produced by bacteria in food.

Medications

Many medications can potentially cause nausea. Some of the most frequently associated include cancer and systemic chemotherapy regimens and general anaesthetic agents.

Pregnancy

Nausea or "morning sickness" is common during early pregnancy but may occasionally continue into the second and third trimesters. In the first trimester nearly 80% of women have some degree of nausea. Pregnancy should therefore be considered as a possible cause of nausea in any women of child bearing age. While usually it is mild and self-limiting, severe cases known as hyperemesis gravidarum may require treatment.

Disequilibrium

A number of conditions involving balance such as motion sickness and vertigo can lead to nausea and vomiting.

Stress and Depression

Nausea may be caused by stress and depression.

Potentially Serious

While most causes of nausea are not serious, some serious causes do occur. These include: Intracranial Pressure secondary to head trauma or hemorrhagic stroke, diabetic ketoacidosis, brain tumor, surgical problems, heart attack, pancreatitis, small bowel obstruction, meningitis, appendicitis, cholecystitis, Addisonian crisis, Choledocholithiasis (from gallstones), hepatitis, as a sign of carbon monoxide poison and many others.

Comprehensive List

Inside the Abdomen

Obstructing Disorders

- Pyloric obstruction
- Small bowel obstruction
- Colonic obstruction
- Superior mesenteric artery syndrome

Enteric Infections

- Viral infection
- Bacterial infection

Inflammatory Diseases

- Cholecystitis
- Pancreatitis
- Appendicitis
- Hepatitis

Sensorimotor Dysfunction

- Gastroparesis
- Intestinal pseudo-obstruction
- Gastroesophageal reflux disease
- Chronic idiopathic nausea
- Functional vomiting
- Cyclic vomiting syndrome
- Rumination syndrome

Other

- Biliary colic
- Abdominal irradiation

Outside the Abdomen

Cardiopulmonary

- Cardiomyopathy

- Myocardial infarction (heart attack)

Inner-ear Diseases

- Motion sickness
- Labyrinthitis
- Malignancy

Intracerebral Disorders

- Malignancy
- Hemorrhage
- Abscess
- Hydrocephalus

Psychiatric Illnesses

- Anorexia and bulimia nervosa
- Depression

Other

- Post-operative vomiting

Medications and Metabolic Disorders

Drugs

- Chemotherapy
- Antibiotics
- Antiarrhythmics
- Digoxin
- Oral hypoglycemic medications
- Oral contraceptives

Endocrine/Metabolic Disease

- Pregnancy
- Uremia
- Ketoacidosis
- Thyroid and parathyroid disease
- Adrenal insufficiency

Toxins

- Liver failure
- Alcohol

Diagnostic Approach

Patient History

Taking a thorough patient history may reveal important clues to the cause of nausea and vomiting. If the patient's symptoms have an acute onset, then drugs, toxins, and infections are likely. In contrast, a long-standing history of nausea will point towards a chronic illness as the culprit. The timing of nausea and vomiting after eating food is an important factor to pay attention to. Symptoms that occur within an hour of eating may indicate an obstruction proximal to the small intestine, such as gastroparesis or pyloric stenosis. An obstruction further down in the intestine or colon will cause delayed vomiting. An infectious cause of nausea and vomiting such as gastroenteritis may present several hours to days after the food was ingested. The contents of the emesis is a valuable clue towards determining the cause. Bits of fecal matter in the emesis indicate obstruction in the distal intestine or the colon. Emesis that is of a bilious nature (greenish in color) localizes the obstruction to a point past the stomach. Emesis of undigested food points to an obstruction prior to the gastric outlet, such as achalasia or Zenker's diverticulum. If patient experiences reduced abdominal pain after vomiting, then obstruction is a likely etiology. However, vomiting does not relieve the pain brought on by pancreatitis or cholecystitis.

Physical Exam

It is important to watch out for signs of dehydration, such as orthostatic hypotension and skin turgor. Auscultation of the abdomen can produce several clues to the cause of nausea and vomiting. A high-pitched tinkling sound indicates possible bowel obstruction, while a splashing "succussion" sound is more indicative of gastric outlet obstruction. Eliciting pain on the abdominal exam when pressing on the patient may indicate an inflammatory process. Signs such as papilledema, visual field losses, or focal neurological deficits are red flag signs for elevated intracranial pressure.

Diagnostic Testing

When a history and physical exam are not enough to determine the cause of nausea and vomiting, certain diagnostic tests may prove useful. A chemistry panel would be useful for electrolyte and metabolic abnormalities. Liver function tests and lipase would identify pancreaticobiliary diseases. Abdominal X-rays showilng air-fluid levels indicate bowel obstruction, while an X-ray showing air-filled bowel loops are more indicative of ileus. More advanced imaging and procedures may be necessary, such as a CT scan, upper endoscopy, colonoscopy, barium enema, or MRI. Abnormal GI motility can be assessed using specific tests like gastric scintigraphy, wireless motility capsules, and small-intestinal manometry.

Pathophysiology

Research on nausea and vomiting has relied on using animal models to mimic the anatomy and neuropharmacologic features of the human body. The physiologic mechanism of nausea is a complex process that has yet to be fully elucidated. There are four general pathways that are activated by specific triggers in the human body that go on to create the sensation of nausea and vomiting.

- Central nervous system (CNS): Stimuli can affect areas of the CNS including the cerebral cortex and the limbic system. These areas are activated by elevated intracranial pressure, irritation of the meninges (i.e. blood or infection), and extreme emotional triggers such as anxiety.
- Chemoreceptor trigger zone (CTZ): The CTZ is located in the area postrema in the floor of the fourth ventricle within the brain. This area is outside the blood brain barrier, and is therefore readily exposed to substances circulating through the blood and cerebral spinal fluid. Common triggers of the CTZ include metabolic abnormalities, toxins, and medications. Activation of the CTZ is mediated by dopamine (D2) receptors, serotonin (5HT3) receptors, and neurokinin receptors (NK1).
- Vestibular system: This system is activated by disturbances to the vestibular apparatus in the inner ear. These include movements that cause motion sickness and dizziness. This pathway is triggered via histamine (H1) receptors and acetylcholine (ACh) receptors.
- Peripheral Pathways: These pathways are triggered via chemoreceptors and mechanoreceptors in the gastrointestinal tract, as well as other organs such as the heart and kidneys. Common activators of these pathways include toxins present in the gastrointestinal lumen and distension of the gastrointestinal lumen from blockage or dysmotility of the bowels. Signals from these pathways travel via multiple neural tracts including the vagus, glossopharyngeal, splanchnic, and sympathetic nerves.

Signals from any of these pathways then travel to the brainstem, activating several structures including the nucleus of the solitary tract, the dorsal motor nucleus of the vagus, and central pattern generator. These structures go on to signal various downstream effects of nausea and vomiting. The body's motor muscle responses involve halting the muscles of the gastrointestinal tract, and in fact causing reversed propulsion of gastric contents towards the mouth while increasing abdominal muscle contraction. Autonomic effects involve increased salivation and the sensation of feeling faint that often occurs with nausea and vomiting.

Treatment

If dehydration is present due to loss of fluids from severe vomiting, rehydration with oral electrolyte solutions is preferred. If this is not effective or possible, intravenous rehydration may be required. Medical care is recommended if: a person cannot keep any liquids down, has symptoms more than 2 days, is weak, has a fever, has stomach pain, vomits more two times in a day or does not urinate for more than 8 hours.

Medications

Many pharmacologic medications are available for the treatment of nausea. There is no medication that is clearly superior to other medications for all cases of nausea. The choice of antiemetic medication may be based on the situation during which the person experiences nausea. For people with motion sickness and vertigo, antihistamines and anticholinergics such as meclizine and scopalamine are particularly effective. Nausea and vomiting associated with migraine headaches respond best to dopamine antagonists such as metoclopramide, prochlorperazine, and chlorpromazine. In cases of gastroenteritis, serotonin antagonist such as ondansetron were found to suppress nausea and vomiting, as well as reduce the need for IV fluid resuscitation. The combination of pyridoxine and doxylamine is the first line treatment for pregnancy-related nausea and vomiting. Dimenhydrinate is an inexpensive and effective over the counter medication for preventing postoperative nausea and vomiting. Other factors to consider when choosing an antiemetic medication include the person's preference, side-effect profile, and cost.

Alternative Medicine

In certain people, cannabinoids may be effective in reducing chemotherapy associated nausea and vomiting. Several studies have demonstrated the therapeutic effects of cannabinoids for nausea and vomiting in the advanced stages of illnesses such as cancer and AIDS.

In hospital settings topical anti-nausea gels are not indicated because of lack of research backing their efficacy. Topical gels containing lorazepam, diphenhydramine, and haloperidol are sometimes used for nausea but are not equivalent to more established therapies.

Ginger has also been shown to be potentially effective in treating several types of nausea.

Tentative evidence supports acupuncture at point PC6.

Prognosis

The outlook depends on the cause. Most people recover within few hours or a day. While short-term nausea and vomiting are generally harmless, they may sometimes indicate a more serious condition. When associated with prolonged vomiting, it may lead to dehydration and/or dangerous electrolyte imbalances. Repeated intentional vomiting, characteristic of bulimia, can cause stomach acid to wear away at the enamel in teeth.

Epidemiology

Nausea and or vomiting is the main complaint in 1.6% of visits to family physicians in Australia. However, only 25% of people with nausea visit their family physician. It is most common in those 15–24 years old and less common in other ages.

Vomiting

Vomiting, also known as emesis and throwing up, among other terms, is the involuntary, forceful expulsion of the contents of one's stomach through the mouth and sometimes the nose.

Vomiting can be caused by a wide variety of conditions; it may present as a specific response to ailments like gastritis or poisoning, or as a non-specific sequela of disorders ranging from brain tumors and elevated intracranial pressure to overexposure to ionizing radiation. The feeling that one is about to vomit is called nausea, which often precedes, but does not always lead to, vomiting. Antiemetics are sometimes necessary to suppress nausea and vomiting. In severe cases, where dehydration develops, intravenous fluid may be required.

Vomiting is different from regurgitation, although the two terms are often used interchangeably. Regurgitation is the return of undigested food back up the esophagus to the mouth, without the force and displeasure associated with vomiting. The causes of vomiting and regurgitation are generally different.

Complications

Aspiration of Vomit

Vomiting can be dangerous if the gastric content enters the respiratory tract. Under normal circumstances the gag reflex and coughing prevent this from occurring; however, these protective reflexes are compromised in persons under the influences of certain substances such as alcohol or anesthesia. The individual may choke and asphyxiate or suffer an aspiration pneumonia.

Dehydration and Electrolyte Imbalance

Prolonged and excessive vomiting depletes the body of water (dehydration), and may alter the electrolyte status. Gastric vomiting leads to the loss of acid (protons) and chloride directly. Combined with the resulting alkaline tide, this leads to hypochloremic metabolic alkalosis (low chloride levels together with high HCO−3 and CO2 and increased blood pH) and often hypokalemia (potassium depletion). The hypokalemia is an indirect result of the kidney compensating for the loss of acid. With the loss of intake of food the individual may eventually become cachectic. A less frequent occurrence results from a vomiting of intestinal contents, including bile acids and HCO−3, which can cause metabolic acidosis.

Mallory–weiss Tear

Repeated or profuse vomiting may cause erosions to the esophagus or small tears in the esophageal mucosa (Mallory–Weiss tear). This may become apparent if fresh red blood is mixed with vomit after several episodes.

Dentistry

Recurrent vomiting, such as observed in bulimia nervosa, may lead to destruction of the tooth enamel due to the acidity of the vomit. Digestive enzymes can also have a negative effect on oral health, by degrading the tissue of the gums.

Pathophysiology

Receptors on the floor of the fourth ventricle of the brain represent a chemoreceptor trigger zone, known as the area postrema, stimulation of which can lead to vomiting. The area postrema is a circumventricular organ and as such lies outside the blood–brain barrier; it can therefore be stimulated by blood-borne drugs that can stimulate vomiting or inhibit it.

14th-century illustration of vomiting from the Casanatense *Tacuinum Sanitatis*.

There are various sources of input to the vomiting center:

- The chemoreceptor trigger zone at the base of the fourth ventricle has numerous dopamine D_2 receptors, serotonin 5-HT_3 receptors, opioid receptors, acetylcholine receptors, and receptors for substance P. Stimulation of different receptors are involved in different pathways leading to emesis, in the final common pathway substance P appears involved.
- The vestibular system, which sends information to the brain via cranial nerve VIII (vestibulocochlear nerve), plays a major role in motion sickness, and is rich in muscarinic receptors and histamine H_1 receptors.
- The cranial nerve X (vagus nerve) is activated when the pharynx is irritated, leading to a gag reflex.
- The vagal and enteric nervous system inputs transmit information regarding the state of the gastrointestinal system. Irritation of the GI mucosa by chemotherapy, radiation, distention, or acute infectious gastroenteritis activates the 5-HT_3 receptors of these inputs.
- The CNS mediates vomiting that arises from psychiatric disorders and stress from higher brain centers.

The vomiting act encompasses three types of outputs initiated by the chemoreceptor trigger zone: Motor, parasympathetic nervous system (PNS), and sympathetic nervous system (SNS). They are as follows:

- Increased salivation to protect tooth enamel from stomach acids. (Excessive vomiting leads to dental erosion). This is part of the PNS output.
- The body takes a deep breath to avoid aspirating vomit.
- Retroperistalsis starts from the middle of the small intestine and sweeps up digestive tract contents into the stomach, through the relaxed pyloric sphincter.

- Intrathoracic pressure lowers (by inspiration against a closed glottis), coupled with an increase in abdominal pressure as the abdominal muscles contract, propels stomach contents into the esophagus as the lower esophageal sphincter relaxes. The stomach itself does not contract in the process of vomiting except for at the angular notch, nor is there any retroperistalsis in the esophagus.
- Vomiting is ordinarily preceded by retching.
- Vomiting also initiates an SNS response causing both sweating and increased heart rate.

The neurotransmitters that regulate vomiting are poorly understood, but inhibitors of dopamine, histamine, and serotonin are all used to suppress vomiting, suggesting that these play a role in the initiation or maintenance of a vomiting cycle. Vasopressin and neurokinin may also participate.

Phases

The vomiting act has two phases. In the retching phase, the abdominal muscles undergo a few rounds of coordinated contractions together with the diaphragm and the muscles used in respiratory inspiration. For this reason, an individual may confuse this phase with an episode of violent hiccups. In this retching phase nothing has yet been expelled. In the next phase, also termed the expulsive phase, intense pressure is formed in the stomach brought about by enormous shifts in both the diaphragm and the abdomen. These shifts are, in essence, vigorous contractions of these muscles that last for extended periods of time - much longer than a normal period of muscular contraction. The pressure is then suddenly released when the upper esophageal sphincter relaxes resulting in the expulsion of gastric contents. Individuals who do not regularly exercise their abdominal muscles may experience pain in those muscles for a few days. The relief of pressure and the release of endorphins into the bloodstream after the expulsion causes the vomiter to feel better.

Contents

Gastric secretions and likewise vomit are highly acidic. Recent food intake appears in the gastric vomit. Irrespective of the content, vomit tends to be malodorous.

The content of the vomitus (vomit) may be of medical interest. Fresh blood in the vomit is termed hematemesis ("blood vomiting"). Altered blood bears resemblance to coffee grounds (as the iron in the blood is oxidized) and, when this matter is identified, the term coffee-ground vomiting is used. Bile can enter the vomit during subsequent heaves due to duodenal contraction if the vomiting is severe. Fecal vomiting is often a consequence of intestinal obstruction or a gastrocolic fistula and is treated as a warning sign of this potentially serious problem (*signum mali ominis*).

If the vomiting reflex continues for an extended period with no appreciable vomitus, the condition is known as non-productive emesis or "dry heaves", which can be painful and debilitating.

Color of Vomit

- Bright red in the vomit suggests bleeding from the esophagus.

- Dark red vomit with liver-like clots suggests profuse bleeding in the stomach, such as from a perforated ulcer.
- Coffee-ground-like vomit suggests less severe bleeding in the stomach, because the gastric acid has had time to change the composition of the blood.
- Yellow vomit suggests bile, indicating that the pyloric valve is open and bile is flowing into the stomach from the duodenum (this is more common in older people).

Differential Diagnosis

Vomiting may be due to a large number of causes, and protracted vomiting has a long differential diagnosis.

Digestive Tract

Causes in the digestive tract:

- Gastritis (inflammation of the gastric wall).
- Gastroenteritis.
- Gastroesophageal reflux disease.
- Pyloric stenosis (in babies, this typically causes a very forceful "projectile vomiting" and is an indication for urgent surgery).
- Bowel obstruction.
- Overeating.
- Acute abdomen and/or peritonitis.
- Ileus.
- Food allergies (often in conjunction with hives or swelling).
- Cholecystitis, pancreatitis, appendicitis, hepatitis.
- Food poisoning.
- In children, it can be caused by an allergic reaction to cow's milk proteins (Milk allergy or lactose intolerance).

Sensory System and Brain

Causes in the sensory system:

- Movement: motion sickness (which is caused by overstimulation of the labyrinthine canals of the ear).
- Ménière's disease.

An intoxicated woman vomiting.

Causes in the brain:

- Concussion.
- Cerebral hemorrhage.
- Migraine.
- Brain tumors, which can cause the chemoreceptors to malfunction.
- Benign intracranial hypertension and hydrocephalus.

Metabolic disturbances (these may irritate both the stomach and the parts of the brain that coordinate vomiting):

- Hypercalcemia (high calcium levels).
- Uremia (urea accumulation, usually due to renal failure).
- Adrenal insufficiency.
- Hypoglycemia.
- Hyperglycemia.

Pregnancy:

- Hyperemesis, morning sickness.

Drug reaction (vomiting may occur as an acute somatic response to):

- Alcohol (being sick while being drunk or being sick the next morning, suffering from the after-effects, i.e., the hangover).
- Opioids.
- Selective serotonin reuptake inhibitors.
- Many chemotherapy drugs.
- Some entheogens (such as peyote or ayahuasca).

Illness (sometimes colloquially known as "stomach flu"—a broad name that refers to gastric inflammation caused by a range of viruses and bacteria):

- Norovirus (formerly Norwalk virus or Norwalk agent).
- Swine influenza.

Emetics

An *emetic*, such as syrup of ipecac, is a substance that induces vomiting when administered orally or by injection. An emetic is used medically when a substance has been ingested and must be expelled from the body immediately (for this reason, many toxic and easily digestible products such as rat poison contain an emetic). Inducing vomiting can remove the substance before it is absorbed into the body. Ipecac abuse can cause detrimental health effects.

Salt water and mustard water have been used since ancient times as emetics. Care must be taken with salt, as excessive intake can potentially be harmful.

Copper sulfate was also used in the past as an emetic. It is now considered too toxic for this use.

Hydrogen peroxide is used as an emetic in veterinary practice.

Social Cues

A drunk man vomiting, while a young slave is holding his forehead. Brygos Painter, 500–470 BC

It is quite common that, when one person vomits, others nearby become nauseated, particularly when smelling the vomit of others, often to the point of vomiting themselves. It is believed that this is an evolved trait among primates. Many primates in the wild tend to browse for food in small groups. Should one member of the party react adversely to some ingested food, it may be advantageous (in a survival sense) for other members of the party to also vomit. This tendency in human populations has been observed at drinking parties, where excessive consumption of alcoholic beverages may cause a number of party members to vomit nearly simultaneously, this being triggered by the initial vomiting of a single member of the party. This phenomenon has been touched on in popular culture: notorious instances appear in the films *Monty Python's The Meaning of Life* (1983) and *Stand By Me* (1986).

Intense vomiting in ayahuasca ceremonies is a common phenomenon. However, people who experience "la purga" after drinking ayahuasca, in general, regard the practice as both a physical and spiritual cleanse and often come to welcome it. It has been suggested that the consistent emetic effects of ayahuasca—in addition to its many other therapeutic properties—was of medicinal benefit to indigenous peoples of the Amazon, in helping to clear parasites from the gastrointestinal system.

There have also been documented cases of a single ill and vomiting individual inadvertently causing others to vomit, when they are especially fearful of also becoming ill, through a form of mass hysteria.

Special bags are often supplied on boats for sick passengers to vomit into.

Most people try to contain their vomit by vomiting into a sink, toilet, or trash can, as vomit is difficult and unpleasant to clean. On airplanes and boats, special bags are supplied for sick passengers to vomit into. A special disposable bag (leakproof, puncture-resistant, odorless) containing absorbent material that solidifies the vomit quickly is also available, making it convenient and safe to store until there is an opportunity to dispose of it conveniently.

People who vomit chronically (e.g., as part of an eating disorder such as bulimia nervosa) may devise various ways to hide this disorder.

An online study of people's responses to "horrible sounds" found vomiting "the most disgusting".

Professor Trevor Cox of the University of Salford's Acoustic Research Centre said that "We are pre-programmed to be repulsed by horrible things such as vomiting, as it is fundamental to staying alive to avoid nasty stuff." It is thought that disgust is triggered by the sound of vomiting to protect those nearby from possibly diseased food.

Miscellanea

- Self-induced:
    - Eating disorders (anorexia nervosa or bulimia nervosa).
    - To eliminate an ingested poison (some poisons should not be vomited as they may be more toxic when inhaled or aspirated; it is better to ask for help before inducing vomiting).
    - Some people who engage in binge drinking induce vomiting to make room in their stomachs for more alcohol consumption.
    - People suffering from nausea may induce vomiting in hopes of feeling better.
- After surgery (postoperative nausea and vomiting).
- Disagreeable sights or disgust, smells or thoughts (such as decayed matter, others' vomit, thinking of vomiting), etc.
- Extreme pain, such as intense headache or myocardial infarction (heart attack).
- Violent emotions.
- Cyclic vomiting syndrome (a poorly understood condition with attacks of vomiting).
- High doses of ionizing radiation sometimes trigger a vomit reflex.
- Violent fits of coughing, hiccups, or asthma.
- Anxiety.
- Depression.
- Overexertion (doing too much strenuous exercise can lead to vomiting shortly afterwards).
- Rumination syndrome, an underdiagnosed and poorly understood disorder that causes sufferers to regurgitate food shortly after ingestion.

Other Types

- Projectile vomiting refers to vomiting that ejects the gastric contents with great force. It is a classic symptom of infantile hypertrophic pyloric stenosis, in which it typically follows feeding and can be so forceful that some material exits through the nose.

Treatment

An antiemetic is a drug that is effective against vomiting and nausea. Antiemetics are typically used to treat motion sickness and the side effects of medications such as opioids and chemotherapy.

Antiemetics act by inhibiting the receptor sites associated with emesis. Hence, anticholinergics, antihistamines, dopamine antagonists, serotonin antagonists, and cannabinoids are used as antiemetics.

Evidence to support the use of antiemetics for nausea and vomiting among adults in the emergency department is poor. It is unclear if any medication is better than another or better than no active treatment.

Epidemiology

Nausea and/or vomiting are the main complaints in 1.6% of visits to family physicians in Australia.

References

- Rousseaux, Wanda Haschek, Matthew Wallig, Colin (2009). Fundamentals of toxicologic pathology (2nd ed.). London: Academic. p. 182. ISBN 9780123704696
- MacLachlan, edited by N. James; Dubovi, Edward J. (2009). Fenner's veterinary virology (4th ed.). Amsterdam: Elsevier Academic Press. p. 399. ISBN 9780123751584
- A. Helms, Richard (2006). Textbook of therapeutics : drug and disease management (8. ed.). Philadelphia, Pa. [u.a.]: Lippincott Williams & Wilkins. p. 2003. ISBN 9780781757348
- Webber, Roger (2009). Communicable disease epidemiology and control : a global perspective (3rd ed.). Wallingford, Oxfordshire: Cabi. p. 79. ISBN 978-1-84593-504-7
- Holland, James F.; Kufe, Donald W.; Weichselbaum, Ralph R.; Pollock, Raphael E.; Frei III, Emil; Gansler, Ted S.; Bast Jr., Robert C. (2003). Cancer medicine (6. [ed.]. ed.). Hamilton, Ontario [u.a.]: Decker. ISBN 9781550092134
- Sleisenger, edited by Mark Feldman, Lawrence S. Friedman, Lawrence J. Brandt; consulting editor, Marvin H. (2009). Sleisenger & Fordtran's gastrointestinal and liver disease pathophysiology, diagnosis, management (PDF) (9th ed.). St. Louis, Mo.: MD Consult. p. 783. ISBN 1-4160-6189-4
- Mandell, Gerald L.; Bennett, John E.; Dolin, Raphael (2004). Mandell's Principles and Practices of Infection Diseases (6th ed.). Churchill Livingstone. ISBN 0-443-06643-4
- Feldman, Mark; Friedman, Lawrence S.; Sleisenger, Marvin H. (2002). Sleisenger & Fordtran's Gastrointestinal and Liver Disease (7th ed.). Saunders. ISBN 0-7216-8973-6
- Charles E. Rosenberg (2009). The Cholera Years the United States in 1832, 1849, and 1866. Chicago: University of Chicago Press. p. 74. ISBN 9780226726762
- Smith, David F.; Diack, H. Lesley and Pennington, T. Hugh (2005) Food Poisoning, Policy and Politics : Corned Beef and Typhoid in Britain in the 1960s. Boydell Press. ISBN 1-84383-138-4
- Shors, Teri (2013). The microbial challenge : a public health perspective (3rd ed.). Burlington, Mass.: Jones & Bartlett Learning. p. 457. ISBN 9781449673338
- Zimmerman, Jeffrey; Karriker, Locke; Ramirez, Alejandro; Kent Schwartz; Gregory Stevenson (15 May 2012). Diseases of Swine (10th ed.). Chichester, West Sussex: John Wiley & Sons. p. 504. ISBN 978-0-8138-2267-9
- Basem Abdelmalak; John Doyle, eds. (2013). Anesthesia for otolaryngologic surgery. Cambridge University Press. pp. 282–287. ISBN 1107018676
- Prober, edited by Sarah Long, Larry Pickering, Charles G. (2012). Principles and practice of pediatric infectious diseases (4th ed.). Edinburgh: Elsevier Saunders. p. 96. ISBN 9781455739851
- Kasper DL, Braunwald E, Fauci AS, Hauser SL, Longo DL, Jameson JL. Harrison's Principles of Internal Medicine. New York: McGraw-Hill, 2005. ISBN 0-07-139140-1

- Dolin, [edited by] Gerald L. Mandell, John E. Bennett, Raphael (2010). "93". Mandell, Douglas, and Bennett's principles and practice of infectious diseases (7th ed.). Philadelphia, PA: Churchill Livingstone/Elsevier. ISBN 0-443-06839-9
- Tintinalli, Judith E. (2010). Emergency Medicine: A Comprehensive Study Guide (Emergency Medicine (Tintinalli)). New York: McGraw-Hill Companies. pp. 830–839. ISBN 0-07-148480-9

Permissions

All chapters in this book are published with permission under the Creative Commons Attribution Share Alike License or equivalent. Every chapter published in this book has been scrutinized by our experts. Their significance has been extensively debated. The topics covered herein carry significant information for a comprehensive understanding. They may even be implemented as practical applications or may be referred to as a beginning point for further studies.

We would like to thank the editorial team for lending their expertise to make the book truly unique. They have played a crucial role in the development of this book. Without their invaluable contributions this book wouldn't have been possible. They have made vital efforts to compile up to date information on the varied aspects of this subject to make this book a valuable addition to the collection of many professionals and students.

This book was conceptualized with the vision of imparting up-to-date and integrated information in this field. To ensure the same, a matchless editorial board was set up. Every individual on the board went through rigorous rounds of assessment to prove their worth. After which they invested a large part of their time researching and compiling the most relevant data for our readers.

The editorial board has been involved in producing this book since its inception. They have spent rigorous hours researching and exploring the diverse topics which have resulted in the successful publishing of this book. They have passed on their knowledge of decades through this book. To expedite this challenging task, the publisher supported the team at every step. A small team of assistant editors was also appointed to further simplify the editing procedure and attain best results for the readers.

Apart from the editorial board, the designing team has also invested a significant amount of their time in understanding the subject and creating the most relevant covers. They scrutinized every image to scout for the most suitable representation of the subject and create an appropriate cover for the book.

The publishing team has been an ardent support to the editorial, designing and production team. Their endless efforts to recruit the best for this project, has resulted in the accomplishment of this book. They are a veteran in the field of academics and their pool of knowledge is as vast as their experience in printing. Their expertise and guidance has proved useful at every step. Their uncompromising quality standards have made this book an exceptional effort. Their encouragement from time to time has been an inspiration for everyone.

The publisher and the editorial board hope that this book will prove to be a valuable piece of knowledge for students, practitioners and scholars across the globe.

Index

www.ingramcontent.com/pod-product-compliance
Lightning Source LLC
LaVergne TN
LVHW080716210926
848485LV00003B/474